Selected Papers on Analysis and Differential Equations

American Mathematical Society

TRANSLATIONS

Series 2 • Volume 211

Selected Papers on Analysis and Differential Equations

Katsumi Nomizu
Editor

American Mathematical Society
Providence, Rhode Island

Translation edited by KATSUMI NOMIZU

2000 *Mathematics Subject Classification.* Primary 00B55;
Secondary 35–XX, 42–XX, 46L35, 58Jxx, 74–XX.

Library of Congress Cataloging-in-Publication Data

Selected papers on analysis and differential equations / Katsumi Nomizu, editor.
p. cm. — (American Mathematical Society translatuins, ISSN 0065-9290 ; ser. 2, v. 211)
Includes bibliographical references.
ISBN 0-8218-3508-4 (acid-free paper)
1. Differential equations, Partial. 2. Fourier analysis. 3. Selfadjoint operators. 4. Differential operators. I. Nomizu, Katsumi, 1924– II. Series.

QA3.A572 ser. 2, vol. 211
[QA374]
510 s—dc22 2003055613
[515′.353]

♾ The paper used in this book is acid-free and falls within the guidelines
established to ensure permanence and durability.
Visit the AMS home page at `http://www.ams.org/`

10 9 8 7 6 5 4 3 2 1 08 07 06 05 04 03

Contents

Preface vii

Van Vleck formula for Wiener integrals and Jacobi fields
NOBUYUKI IKEDA 1

Spectral geometry for Schrödinger operators in a magnetic field focusing on geometrical and dynamical structures on manifolds
RUISHI KUWABARA 25

Symbolic dynamics and C^*-algebras
KENGO MATSUMOTO 47

Inverse problems for elasticity
GEN NAKAMURA 71

Time-global solutions of nonlinear evolution equations and their stability
YOSHIHIRO SHIBATA 87

Wavelets and eigenvalues of Schrödinger operators
KAZUYA TACHIZAWA 107

Recent topics on nonlinear partial differential equations: Structure of radial solutions for semilinear elliptic equations
EIJI YANAGIDA AND SHOJI YOTSUTANI 121

Amer. Math. Soc. Transl.
(2) Vol. **211**, 2003

Preface

This is a collection of several papers that originally appeared in the journal *Sūgaku* in Japanese. These translated articles would normally appear in the AMS journal *Sugaku Expositions*. In order to expedite publication, the AMS has chosen, with the consent of the Mathematical Society of Japan, to publish them as a volume of selected papers in the *Translations Series* 2 of the AMS. To give the reader a quick idea of the contents, we give brief quotations from the introduction to each article, with occasional emendations.

The first paper is *Van Vleck formula for Wiener integrals and Jacobi fields*, by Nobuyuki Ikeda. This survey is based on a lecture presented at the annual meeting of the Mathematical Society of Japan in 1996, which in turn was based on joint work with Shigeo Kusuoka and Shojiro Manabe. The key words and phrases include: M. Kac, Feynman path integrals, Levy's formula, Gaussian integrals, Van Vleck formula, Jacobian of a transform, and Selberg's trace formula.

Ruishi Kuwabara wrote on *Spectral geometry for Schrödinger operators in a magnetic field focusing on geometrical and dynamical structures on manifolds*. The aim of this article is to present some topics and results on the relationships among the spectrum of $\hat{H}$, geometry of (M, g, A), and mechanics in $\mathcal{H}_{cl}$ in the case where M is a compact manifold without boundary and the scalar potential $V = 0$. In the framework of symplectic structures and their quantized objects we will consider how the magnetic field Θ or its vector potential A affects the quantum spectrum from the geometric and dynamical viewpoints. The techniques used here are based on the theory of Hamiltonian mechanics, the geometry of connections, and the theory of pseudodifferential and Fourier integral operators. The main results in this survey were obtained around 1990, and we cannot cover the many arguments that have recently been deeply and widely developed om this subject.

In Kengo Matsymoto's paper, *Symbolic dynamics and $C*$-algebras*, the author will deal with two different types of objects. One is discrete dynamical systems, consisting of the shifts on bi-infinite sequences of finite symbols, whose theory is mainly part of topological dynamical systems. The other one is operator algebras, consisting of infinite dimensional algebras of mutually noncommutative operators, whose theory is mainly part of functional analysis. These two objects look very different. In this paper, the author will introduce an interesting relationship between them. Many concepts and interesting applications of the two subjects will indeed come into play; for example, we mention how Hadamard's old ideas were instrumental to the origin of symbolic dynamics.

In *Inverse problems for elasticity*, Gen Nakamura considers such problems with and without residual stress. The inverse problem with residual stress may not be very familiar, but it is very interesting and important in applications. In addition to covering known results, including his own, the author points out some of the open problems, and makes conjectures on possible future results.

Yoshihiro Shibata contributes a paper on *Time-global solutions of nonlinear evolution equations and their stability.* In it he considers a certain problem on stability of stationary solutions to evolution equations, concentrating on his own recent results. The equation to be treated is

$$u_t + A(u) = f,$$

where the subscript t denotes partial differentiation with respect to time.

Kazuya Tachizawa writes on *Wavelets and eigenvalues of Schrödinger operators.* He starts by explaining the basics of wavelet theory, and goes on to consider their applications to eigenvalues of Schrödinger operators. One of the main topics of Fourier analysis is the decomposition of a function; a typical example is the Fourier series decomposition of a periodic function. In such a decomposition it is important to investigate the relation between the norm of the function and its components.

Finally, Eiji Yanagida and Shoji Yotsutani write on *Recent topics in nonlinear partial differential equations: Structure of radial solutions for semilinear elliptic equations.* Over the past two decades there has been a considerable development in the theory of nonlinear PDEs, and many interesting facts have been discovered for various equations. In this article the authors focus on the structure of radial solutions for semilinear elliptic equations with power semilinearity, and related topics, and try to survey the recent progress as clearly as possible.

Amer. Math. Soc. Transl.
(2) Vol. **211**, 2003

Van Vleck Formula for Wiener Integrals and Jacobi Fields

Nobuyuki Ikeda

1. Introduction

A propagator can be written down by using elementary functions in the cases of potentials relative to harmonic oscillators and uniform magnetic fields. Its representation plays a fundamental role in the study of semi-classical approximation and in many problems in wave mechanics.

We obtain this representation using the general formula that Van Vleck [**54**] first proved in 1928 for fundamental solutions of Schödinger equations with quadratic potential.

According to Feynman [**12**], this means that the path integral for the propagator can be represented explicitly. In fact, in 1951, C. Morette [**45**] proved the Van Vleck formula directly for Feynman path integrals. Moreover, this formula contained only so-called classical dynamical quantities associated with the classical paths (critical points of the action integral constructed from Lagrangians) and Jacobi fields along their trajectories. In this sense, the representation derived from the Van Vleck formula takes a special position among many analogous path integral formulas [15].

On the other hand, we can obtain an explicit expression for fundamental solutions of heat equations with the above mentioned potentials. This expression is applicable in the theory of partial differential equations and functional analysis. As one can immediately recognize, this expression looks like that for propagators. Furthermore, since those fundamental solutions have an integral representation using the (generalized) Wiener integral, by the method of stochastic analysis we can consider this expression as an expression for the Wiener integral. These formulae can be proved by a direct calculation or by the method of eigenfunction expansion. They can also be proved by at least the following two methods. The one is a transformation method in Wiener space studied by R. H. Cameron and W. T. Martin, the other is the method of Fourier expansion of Brownian motion. On the other hand, these methods also play an important role in the study of Feynman path integrals. But few mathematicians, to the author's knowledge, except M. Kac [29], [30] and Molchanov [44], have stressed this viewpoint via the Van Vleck formula

This article originally appeared in Japanese in Sūgaku **52** (3) (2000), 225–244.
2000 *Mathematics Subject Classification.* Primary 58J65; Secondary 60H30.

in the case of the Wiener integral. In the nineties, related to the study of the Malliavin calculus, a large number of publications on the Fourier transform of the probability distribution of quadratic Wiener functionals have appeared, and the classical dynamical meaning of their representations has been discussed ([6], [48]).

On the other hand, M. Kac has imported thoughts used in the study of Feynman path integrals into the study of Wiener integrals, without attempting to make it mathematically rigorous. The objective of this review, after his idea, is to clarify in what forms the analogues to the Van Vleck formula for the path integral also hold in Wiener space.

The present survey article is based on a lecture presented at the annual meeting of the Mathematical Society of Japan in 1996, which in turn was based on joint work with Shigeo Kusuoka and Shojiro Manabe [19]-[22].

This review consists of two parts. In the first part we mainly take the case of a potential associated with a uniform magnetic field. In this rather special case, by dealing in a unified way with results obtained in various ways, we clarify the method of generalization. As a result, roughly speaking, the core of stochastic analysis concerning the Van Vleck formula is a change of variables of Gaussian integrals in the path space, and it turns out that the central task is how to transform a quadratic form on the path space and to calculate the Jacobian of the transform. Similar problems in finite dimension are already settled, and the results have been applied in many fields of mathematics (for example, see [18], Chapter 7). In Section 2, we explain prerequisites as little as possible, although some prerequisite knowledge of the Wiener measure is needed. If necessary, consult, for example, [33].

In the latter half of the review, in Section 3, following the method employed in [19], we present a framework by which we can systematically formulate the idea described in Section 2. Next, in Sections 4 and 5, we consider typical cases that can be treated in this framework. The so-called generalized Lévy formula, motivated by the study of polymers, is a typical example which can be treated by the method described in Section 4 ([1], [6], [20], [48]). We cover only the case of two-dimensional Euclidean space $\mathbf{R}^2$, in order to avoid complex notation and explanations. In Section 6, we take up the cases (not the case of the Wiener integral itself) where related characteristics can be represented by the quantities of classical dynamics. As a typical example, following Ikeda and Matsumoto [23], we explain the idea, due to M.G. Gutzwiller and others ([16, 17, 8]), about the Selberg trace formula, in which they used the Van Vleck approximation.

In the last section, we make supplementary explanations about a few related topics.

As is well known, defining Feynman path integrals in a mathematically rigorous manner for a rather general class of integrands remains a major problem. But for the simple cases treated in this review much mathematically rigorous research has been done (see, for example, Fujiwara [14], etc.).

2. Case of a uniform magnetic field

We take a potential with uniform magnetic field as a simple example to illustrate known facts from various points of view. This serves as an outline of the later sections.

2.1. Heat equation with uniform magnetic field. Let α and θ be differential one-forms on $\mathbb{R}^2$:

$$\alpha = \lambda\theta, \quad \theta = \frac{1}{2}\big(x^1 dx^2 - x^2 dx^1\big), \qquad x = (x^1, x^2) \in \mathbb{R}^2,$$

where λ is a fixed positive constant. The uniform magnetic field $d\alpha = \lambda dx^1 \wedge dx^2$ determined by α is often identified with λ and is called a uniform magnetic field λ. In this situation,

$$\frac{\partial u}{\partial t} = \frac{1}{2}\big[\Delta u + 2\sqrt{-1}\,\lambda\langle\!\langle du, \theta\rangle\!\rangle - \lambda^2\|\theta\|^2 u\big], \quad (t, x) \in (0,\infty)\times\mathbb{R}^2, \tag{2.1}$$

is called a heat equation with uniform magnetic field $d\alpha$ (or vector potential α). Here $\langle\!\langle \cdot, * \rangle\!\rangle$ is an inner product with respect to the flat metric in the space $\Lambda^1(\mathbb{R}^2)$ of differentiable one-forms on $\mathbb{R}^2$, and $\|\cdot\|^2 = \langle\!\langle \cdot, \cdot\rangle\!\rangle$. The fundamental solution (with respect to the Lebesgue measure) $p(t, x, y)$, $(t, x, y) \in (0, \infty) \times \mathbb{R}^2 \times \mathbb{R}^2$, of the equation (2.1) can be written as ([**20, 24**])

$$\begin{aligned} p(t,x,y) &= \frac{1}{2\pi t}\frac{\lambda t/2}{\sinh(\lambda t/2)}\exp\Big[\sqrt{-1}\,\lambda\psi(x,y) \\ &\qquad - \frac{\lambda t}{2}\coth\Big(\frac{\lambda t}{2}\Big)\frac{(x^1-y^1)^2+(x^2-y^2)^2}{2t}\Big], \\ \psi(x,y) &= \frac{1}{2}\big[x^1y^2 - x^2y^1\big], \qquad x=(x^1,x^2),\ y=(y^1,y^2)\in\mathbb{R}^2. \end{aligned} \tag{2.2}$$

In order to change our standpoint, we consider a two-dimensional Wiener space $\{W, P\}$; let $W = \{w; w \in C([0,t] \to \mathbb{R}^2),\ w(0) = 0\}$, $\mathcal{B}(W)$ the topological σ-field, and P the 2-dimensional Wiener measure on $\{W, \mathcal{B}(W)\}$ ([**24, 33**]). For any $w \in W$ and $x \in \mathbb{R}^2$, we define $w_x \in C([0,t] \to \mathbb{R}^2)$ by $w_x(s) = x + w(s)$, $s \in [0,t]$. A stochastic line integral of a differential one-form θ along a continuous curve $w_x[0,t] = \{w_x(s); s \in [0,t]\}$ is given by $A(t; w_x)$ ([**24**]):

$$A(t;w_x) = \frac{1}{2}\int_0^t \{w_x^1(s)dw_x^2(s) - w_x^2(s)dw_x^1(s)\}, \quad w_x = (w_x^1, w_x^2),$$

and is called a stochastic area of the curve $w_x[0,t]$. Although almost all $w \in W$ are not smooth under the Wiener measure, we can ensure that the property holds for smooth curves even in case of the stochastic version if we define it by Stratonovich's stochastic integral. So hereafter we don't mention that point ([**24**]). If the stochastic line integral is thus defined, it is well-known that

$$\begin{aligned} p(t,x,y) &= \int_W \exp\big[\sqrt{-1}\,\lambda A(t;w_x)\big]\delta_y(w_x(t))P(dw), \\ &\qquad\qquad (t,x,y)\in(0,\infty)\times\mathbb{R}^2\times\mathbb{R}^2 \end{aligned} \tag{2.3}$$

(see for example, [**24, 55**]), where $\delta_y(w_x(t))$ is a composition of Dirac's δ-function δ_y and the nondegenerate Wiener functional $w_x(t)$ in the sense of Malliavin [**35, 36**], and the right-hand side of (2.3) is a *generalized Wiener integral* in the sense of Watanabe [**55**]. To give an accurate meaning to these, one needs some knowledge of stochastic analysis, but this is not an obstacle for understanding the story line, since a generalized integral has almost the same properties as the usual integral. So, we will not enter into the details, which can be found for example in [**24, 55**].

Combining (2.2) and (2.3), we have

$$\begin{aligned}
&\int_W \exp\big[\sqrt{-1}\,\lambda A(t;w_x)\big]\delta_y(w_x(t))P(dw)\\
&\quad = \frac{1}{2\pi t}\frac{\lambda t/2}{\sinh(\lambda t/2)}\\
&\qquad \times \exp\left[\sqrt{-1}\,\lambda\psi(x,y) - \frac{\lambda t}{2}\coth\Big(\frac{\lambda t}{2}\Big)\frac{(x^1-y^1)^2+(x^2-y^2)^2}{2t}\right],\\
&\qquad\qquad x=(x^1,x^2),\ y=(y^1,y^2)\in\mathbb{R}^2.
\end{aligned}\tag{2.4}$$

This equality is a paraphrase of Lévy's stochastic area formula in terms of the generalized Wiener integral ([**20, 24, 34**]).

2.2. Van Vleck formula for a propagator. In this subsection, in order to grasp the meaning of the equation (2.4), we digress to recapitulate some known results in the theory of the Schrödinger equation and the Feynman path integral.

We consider a charged particle with electric charge e and mass m under a uniform magnetic field λ. Its Lagrangian is

$$\begin{aligned}
L(x,\dot{x}) &= \frac{m}{2}\big((\dot{x}^1)^2+(\dot{x}^2)^2\big)+\frac{e\lambda}{2c}(x^1\dot{x}^2-x^2\dot{x}^1),\\
&\qquad x=(x^1,x^2),\ \dot{x}=(\dot{x}^1,\dot{x}^2)
\end{aligned}\tag{2.5}$$

([**12, 32**]). Here c is the speed of light. The corresponding Schrödinger equation is

$$\begin{aligned}
\frac{\hbar}{\sqrt{-1}}\frac{\partial u}{\partial t} &= \frac{1}{2m}\left[\hbar^2\Delta u - 2\sqrt{-1}\,\hbar\,\frac{e\lambda}{c}\langle\langle du,\theta\rangle\rangle - \Big(\frac{e\lambda}{c}\Big)^2\|\theta\|^2 u\right],\\
&\qquad (t,x)\in(0,\infty)\times\mathbb{R}^2,
\end{aligned}\tag{2.6}$$

where $\hbar$ is Planck's constant divided by 2π. Then the propagator $K(t,x,y)$, $(t,x,y)\in(0,\infty)\times\mathbb{R}^2\times\mathbb{R}^2$, describing the quantum mechanical state in wave mechanics, is obtained as a fundamental solution of this equation. Moreover, if $0<\omega t<\pi$, then

$$\begin{aligned}
K(t,x,y) = \frac{m}{\hbar\sqrt{-1}}\frac{1}{2\pi t}\frac{\omega t/2}{\sin(\omega t/2)}\exp&\left[\frac{\sqrt{-1}\,m\omega}{\hbar}\psi(x,y)\right.\\
&\left.+\frac{m\sqrt{-1}}{\hbar}\frac{\omega t}{2}\cot\Big(\frac{\omega t}{2}\Big)\frac{(x^1-y^1)^2+(x^2-y^2)^2}{2t}\right],\\
x=(x^1,x^2),\ y=(y^1,y^2)&\in\mathbb{R}^2
\end{aligned}\tag{2.7}$$

where $\omega = e\lambda/mc$ (for example, see [**12, 32, 50**]). Though there are many techniques, including direct calculation, that lead to this fact, here we pay attention to the Van Vleck formula, which holds not only in this case, but in wider cases. For $t>0$, $x,y\in\mathbb{R}^2$, we set

$$\Omega(t,x,y)=\left\{\phi\ ;\ \begin{array}{l}\phi\in C([0,t]\to\mathbb{R}^2),\ \phi(0)=x,\ \phi(t)=y \text{ and each component}\\ \phi^i(s),\ 0\leqq s\leqq t,\ i=1,2, \text{ of } \phi(s)=(\phi^1(s),\phi^2(s)) \text{ is abso-}\\ \text{lutely continuous and has a square integrable differential}\end{array}\right\}.$$

We consider a class of Lagrangians $L(x,\dot{x})$ that are quadratic forms in x, $\dot{x}$. We introduce an action integral

$$S(t,x,y;\phi)=\int_0^t L(\phi(s),\dot{\phi}(s))ds,\qquad \phi\in\Omega(t,x,y),\tag{2.8}$$

where $\dot{\phi}$ is the differentiation of ϕ with respect to t. We *assume* that there exists a unique classical trajectory, that is, a critical point of the functional $S(t,x,y;\phi)$ on $\Omega(t,x,y)$, and that $S(t,x,y;\cdot)$ is nondegenerate at $\phi_{\rm cl}$. We then set $S_{\rm cl}(t,x,y) = S(t,x,y;\phi_{\rm cl})$. Next, we consider a matrix $\left(\partial^2 S_{\rm cl}/\partial x \partial y\right)$ whose (i,j) component is given by

$$\frac{\partial^2 S_{\rm cl}(t,x,y)}{\partial x^i \partial y^j}, \qquad i,j = 1,2, \qquad x = (x^1,x^2),\ y = (y^1,y^2) \in \mathbb{R}^2.$$

This is called a *Van Vleck matrix.* The following relation holds for $K(t,x,y)$ corresponding to the quadratic Lagrangian $L(x,\dot{x})$ ([**9, 10, 29, 50, 54**]):
Van Vleck formula ([**54**])

$$K(t,x,y) = \frac{1}{2\pi\hbar\sqrt{-1}}\left\{\det\left(\frac{\partial^2 S_{\rm cl}}{\partial x \partial y}\right)\right\}^{1/2} \exp\left[\frac{\sqrt{-1}}{\hbar} S_{\rm cl}(t,x,y)\right]. \tag{2.9}$$

Equation (2.7) can be obtained as follows. For the Lagrangian L given by (2.5), if the condition $0 < \omega t < \pi$ holds then the assumption mentioned above is satisfied, and further, since

$$S_{\rm cl}(t,x,y) = m\left\{\omega\psi(x,y) + \frac{\omega t}{2}\cot\left(\frac{\omega t}{2}\right)\frac{(x^1-y^1)^2 + (x^2-y^2)^2}{2t}\right\}, \tag{2.10}$$

substituting this into (2.9), we obtain (2.7).

Next we present a Van Vleck formula of a different form, due to C. Morette [**45**]. Since, using a Feynman path integral, the propagator $K(t,x,y)$ can be rewritten as

$$K(t,x,y) = \int_{\Omega(t,x,y)} \exp\left[\frac{\sqrt{-1}}{\hbar}\, S(t,x,y;\phi)\right]\mathcal{D}(\phi), \tag{2.11}$$

the Van Vleck formula can be written in the following form:

$$\begin{aligned}&\int_{\Omega(t,x,y)} \exp\left[\frac{\sqrt{-1}}{\hbar}\, S(t,x,y;\phi)\right]\mathcal{D}(\phi)\\ &\qquad = \frac{1}{2\pi\hbar\sqrt{-1}}\left\{\det\left(\frac{\partial^2 S_{\rm cl}}{\partial x \partial y}\right)\right\}^{1/2} \exp\left[\frac{\sqrt{-1}}{\hbar} S_{\rm cl}(t,x,y)\right]\end{aligned} \tag{2.12}$$

([**32, 45, 50**]).

2.3. Van Vleck formula for a fundamental solution of the heat equation. In the case of a uniform magnetic field, comparing the methods for the propagator and for the heat equation, we see that the equations (2.1) and (2.6) and the representations (2.2) and (2.7) are transformed into each other by a change of parameters. Moreover, rewriting the form of Wiener measure in the (informal) Feynman path integral style, we see that the right-hand side of (2.3) corresponds to that of (2.11). Therefore in the case of the heat equation, that is, the Wiener integral, we can expect that identities analogous to the Van Vleck formulas (2.9) and (2.12) also hold. So we take up a quadratic form analogous to the Lagrangian in (2.5):

$$\begin{aligned}L(x,\dot{x}) = \frac{1}{2}\left((\dot{x}^1)^2 + (\dot{x}^2)^2\right) - \frac{\sqrt{-1}\,\lambda}{2}\,(x^1\dot{x}^2 - x^2\dot{x}^1),&\\ x = (x^1,x^2),\ \dot{x} = (\dot{x}^1,\dot{x}^2),&\end{aligned} \tag{2.13}$$

and we define $\Omega^*(t,x,y)$ by replacing $C([0,t] \to \mathbb{R}^2)$ by $C([0,t] \to \mathbb{C}^2)$ in the definition of $\Omega(t,x,y)$.

For L, we define $S(t,x,y;\phi)$ on $\Omega^*(t,x,y)$ by (2.8). Then its critical point ϕ_{cl} is uniquely determined, and $S(t,x,y;\phi)$ is a nondegenerate quadratic form at ϕ_{cl}. Setting $S_{\mathrm{cl}}(t,x,y) = S(t,x,y;\phi_{\mathrm{cl}})$, we have

$$S_{\mathrm{cl}}(t,x,y) = -\sqrt{-1}\,\lambda\psi(x,y) + \frac{\lambda t}{2}\coth\Big(\frac{\lambda t}{2}\Big)\frac{(x^1-y^1)^2+(x^2-y^2)^2}{2t}. \tag{2.14}$$

Combining this with (2.2) and (2.4), we obtain

$$\begin{aligned} p(t,x,y) &= \int_W \exp\big[\sqrt{-1}\,\lambda A(t;w_x)\big]\delta_y(w_x(t))P(dw) \\ &= \frac{1}{2\pi}\Big\{\det\Big(\frac{\partial^2 S_{\mathrm{cl}}}{\partial x \partial y}\Big)\Big\}^{1/2} \exp\big[-S_{\mathrm{cl}}(t,x,y)\big]. \end{aligned} \tag{2.15}$$

Summarizing these, we can state our first objective as follows: *If we replace a phase function $A(t;w_x)$ in (2.3) by the more general quadratic form $F(t;w_x)$, after the method for the case of propagator in Subsection 2.2, we can consider a generalization of $S(t,x,y;\phi)$ and can show an analogue of the Van Vleck formula (2.12) corresponding to (2.15) as a general theory, without using a concrete expression such as (2.14).*

2.4. Van Vleck determiant and Fourier expansion. The next issue is to take up the determinant $\det\big(\partial^2 S_{\mathrm{cl}}/\partial x\partial y\big)$ and to consider it from another point of view. For simplicity, in this subsection we always restrict ourselves to the case $t=1$. The general case can easily be reduced to this case. Using the

$$\begin{aligned} &\int_W \exp\big[\sqrt{-1}\,\lambda A(1;w_x)\big]\delta_y(w_x(1))P(dw) \\ &\qquad = \int_W \exp\big[\sqrt{-1}\,\lambda A(1;w)\big]\delta_0(w(1))P(dw)\exp\big[-S_{\mathrm{cl}}(1,x,y)\big] \end{aligned}$$

([**3, 24**]). Next, defining a system of kernels of integral operators $\big\{F^{ij}(s,u)\big\}_{i,j=1,2}$ by

$$\begin{aligned} &F^{11}(s,u) = F^{22}(s,u) = 0, \\ &F^{12}(s,u) = F^{21}(u,s) = \begin{cases} 1/4, & 0 \leqq s < u \leqq 1, \\ -1/4, & 0 \leqq u < s \leqq 1, \end{cases} \end{aligned} \tag{2.16}$$

we see that

$$A(1;w) = \sum_{j,k=1}^{2}\Big[\int_0^1\int_0^s F^{jk}(s,u)dw^k(u)dw^j(s) + \int_0^1\int_0^u F^{jk}(s,u)dw^j(s)dw^k(u)\Big].$$

On the other hand, let H be the Cameron-Martin subspace of W, that is,

$$H = \left\{ h\ ;\ \begin{array}{l} h \in W, \text{ each component } h^i(s),\ 0 \leqq s \leqq t,\ i=1,2, \text{ of } h(s) = \\ (h^1(s),h^2(s)) \text{ is absolutely continuous and its derivative is} \\ \text{square integrable} \end{array} \right\}.$$

This space is a Hilbert space with respect to the following inner product:

$$\langle h_1, h_2\rangle = \sum_{j=1}^{2}\int_0^1 \dot h_1^j(s)\dot h_2^j(s)ds.$$

We then define an operator $B \in \mathrm{HS}^{(s)}(H)$ by

$$B[h](s) = (B[h]^{(1)}(s), B[h]^{(2)}(s)) \in H,$$

$$\frac{d}{ds}B[h]^{(j)}(s) = \sum_{k=1}^{2}\int_0^1 F^{jk}(s,u)\dot{h}^k(u)du, \quad h \in H,\ j = 1,2, \tag{2.17}$$

where $\mathrm{HS}^{(s)}(H)$ means the space of all symmetric Hilbert-Schmidt operators on H. In the following, without mention we consider the complexification of H if necessary and extend operators on H accordingly. For simplicity, to denote the extended operators we use the same notation as before. We sometimes use the notation $A(1;w) = \langle Bw, w\rangle$, since $A(1;w)$ is determined uniquely by $B \in \mathrm{HS}^{(s)}(H)$. Let $H^0 = \{h; h \in H, h(1) = 0\}$ and set $B^{\#} = P_{H^0}BP_{H^0}$. P_{H^0} denotes the orthogonal projection on H^0. In this notation, if we denote

$$J = \begin{pmatrix} 0 & -1 \\ 1 & 0 \end{pmatrix}, \quad \mathcal{I}[h] = (\mathcal{I}[h^1], \mathcal{I}[h^2]), \quad \mathcal{I}[h^i](s) = \int_0^s h^i(u)du,$$
$$h = (h^1, h^2) \in H,$$

then we have

$$\begin{aligned} &B = B_0 + B_1, \quad B^{\#} = B_0 + B_1^{\#}, \\ &\text{where } B_0[h](s) = \frac{1}{2}\mathcal{I}[{}^t(J[{}^th])](s), \quad B_1[h](s) = -\frac{1}{4}{}^t(J[{}^th](1))\, s, \\ &B_1^{\#}[h](s) = -\big(P_{(H^0)^\perp}B_0[h]\big)(s) = -\frac{1}{2}\mathcal{I}[{}^t(J[{}^th])](1)s, \end{aligned} \tag{2.18}$$

and ${}^tv$ means the transposed vector v. This implies that the eigenvalues $\{\lambda_n\}$ of $B^{\#}$ in H^0 are

$$\lambda_n = \frac{1}{2}\frac{1}{2n\pi}, \quad n \in \mathbb{Z}^* = \mathbb{Z}\setminus\{0\}, \tag{2.19}$$

and that all eigenvalues have multiplicity 2. The corresponding eigenfunctions are given by

$$\begin{aligned} \phi_n(s) &= \frac{1}{2n\pi}(\cos 2n\pi s - 1, \sin 2n\pi s), \\ \psi_n(s) &= \frac{1}{2n\pi}(\sin 2n\pi s, -(\cos 2n\pi s - 1)), \end{aligned} \qquad 0 \leqq s \leqq 1,\ n \in \mathbb{Z}^*$$

([**20, 21**]). We obtain a Fourier expansion of Brownian motion: for $w \in W$ a.s.(P)

$$\begin{aligned} w(s) &= \langle \phi_{0,1}, w\rangle\phi_{0,1}(s) + \langle \phi_{0,2}, w\rangle\phi_{0,2}(s) \\ &\quad + \sum_{n\in\mathbb{Z}^*}\langle \phi_n, w\rangle\phi_n(s) + \sum_{n\in\mathbb{Z}^*}\langle \psi_n, w\rangle\psi_n(s), \quad 0 \leqq s \leqq 1, \end{aligned} \tag{2.20}$$

where

$$\begin{aligned} &\phi_{0,1}(s) = (1,0)s, \quad \phi_{0,2}(s) = (0,1)s, \qquad 0 \leqq s \leqq 1, \\ &\langle h, w\rangle = \sum_{k=1}^{2}\int_0^1 \dot{h}^k(s)dw^k(s), \qquad h = (h^1(s), h^2(s)) \in H \end{aligned} \tag{2.21}$$

([**26, 46**]). Since $\big\{(\langle \phi_n, w\rangle, \langle \psi_n, w\rangle), n \in \mathbb{Z}^*\big\}$ is a sequence of independent random variables obeying the two-dimensional standard Gaussian distribution, by (2.20) it follows that *the infinite dimensional Wiener space can be represented as a direct* *product of a countable number of two-dimensional standard Gaussian spaces.* Using this fact, we can transform a quadratic form on W into its canonical form. Next, making a change of variable in the two-dimensional Gaussian integral and then calculating its Jacobian, we have

$$\begin{aligned}
&\int_W \exp\big[\sqrt{-1}\,\lambda A(1;w)\big]\delta_0(w(1))P(dw) \\
&\quad= \frac{1}{2\pi}\prod_{n\in\mathbb{Z}^*}\int_{\mathbb{R}^2}\frac{1}{2\pi}\exp\Big[-\frac{(1-2\sqrt{-1}\,\lambda\lambda_n)(a^2+b^2)}{2}\Big]\,dadb \\
&\quad= \frac{1}{2\pi}\Big(\prod_{n\in\mathbb{Z}^*}(1-2\sqrt{-1}\,\lambda\lambda_n)\Big)^{-1} \\
&\quad= \frac{1}{2\pi}\Big(\prod_{n\in\mathbb{Z}^*}(1-2\sqrt{-1}\,\lambda\lambda_n)\exp\big[2\sqrt{-1}\,\lambda\lambda_n\big]\Big)^{-1} \\
&\quad= \frac{1}{2\pi}\big(\det{}_2(I-2\sqrt{-1}\,\lambda B^{\#})\big)^{-1/2}
\end{aligned}\tag{2.22}$$

($\det_2$ denotes regularized determinant).

See [**22**] to observe how to determine the power 1/2 in the last line of (2.22), Combining this with (2.15), we have

$$\det{}_2(I-2\sqrt{-1}\,\lambda B^{\#}) = \Big\{\det\Big(\frac{\partial^2 S_{\mathrm{cl}}}{\partial x\partial y}\Big)\Big\}^{-1}. \tag{2.23}$$

In short, *the Van Vleck formula gives a representation of the infinite dimensional determinant* $\det_2(I-2\sqrt{-1}\,\lambda B^{\#})$ *in terms of a finite dimensional determinant.* Roughly speaking, this means that only the finite dimensional part of the Jacobian affects a change of variables.

Our next objective is to show in what range (2.23) holds *without using concrete expressions* such as (2.16), (2.17), (2.18), and (2.19).

REMARK 2.1. The idea of reducing an infinite dimensional integral on the path space to an infinite product of finite dimensional Gaussian integrals by using the Fourier expansion was first introduced by Paley and Wiener [**46**], in the case of Wiener space, and was used to calculate some Wiener integrals. Their idea is widely used even now in stochastic calculus. The same idea has been used for the Feynman path integral, since the beginning of the theory of path integrals by Feynman himself (see [**12**], chap. 3, section 3-11).

2.5. Decomposition of Hilbert-Schmidt operators and the Van Vleck formula. Taking account of the decomposition (2.18) and using the idea developed in [**19, 20**], the fact stated in the last part of the preceding subsection can be shown directly. The following equalities can be easily shown by formal calculations:

$$\begin{aligned}
&I-2\sqrt{-1}\,\lambda B^{\#} = Q(\sqrt{-1}\,\lambda)(I-2\sqrt{-1}\,\lambda B_0),\\
&Q(\sqrt{-1}\,\lambda) = I-2\sqrt{-1}\,\lambda B_1^{\#}\,(I-2\sqrt{-1}\,\lambda B_0)^{-1}.
\end{aligned}\tag{2.24}$$

Using the fact that B_0 is an operator of Volterra type and the range of $B_1^{\#}$ is two-dimensional, we have

$$\begin{aligned} \det\nolimits_2(I - 2\sqrt{-1}\,\lambda B^{\#}) &= \det Q(\sqrt{-1}\,\lambda)\exp\bigl[2\sqrt{-1}\,\lambda \operatorname{tr} B_1^{\#}\bigr] \\ &= \det Q(\sqrt{-1}\,\lambda). \end{aligned} \tag{2.25}$$

Further, taking $\phi_{0,1}$, $\phi_{0,2}$ in (2.21) as a basis of the range $(H^0)^{\perp}$ of $B_1^{\#}$ and using the representation of B_0 and $B_1^{\#}$ in (2.18), we easily have the matrix representation of $Q(\sqrt{-1}\,\lambda)$:

$$\begin{aligned} Q(\sqrt{-1}\,\lambda) &= \frac{1}{\sqrt{-1}\,\lambda}\begin{pmatrix} \sqrt{-1}\sinh\lambda & 1-\cosh\lambda \\ -(1-\cosh\lambda) & \sqrt{-1}\sinh\lambda \end{pmatrix}, \\ \det Q(\sqrt{-1}\,\lambda) &= \left\{\frac{\sinh(\lambda/2)}{\lambda/2}\right\}^2 \end{aligned} \tag{2.26}$$

([**19, 20, 22**]). The following equality also holds:

$$\left\{\det\left(\frac{\partial^2 S_{\mathrm{cl}}}{\partial x \partial y}\right)\right\}^{-1} = \det Q(\sqrt{-1}\,\lambda). \tag{2.27}$$

2.6. Jacobi fields and Van Vleck matrix. As Jacobi has shown, the Jacobi fields along classical trajectories are generated by vector fields obtained by partial differentiation of the starting and end points of classical trajectories ([**27, 43**]). This fact also holds in our framework. Using this fact, the operator $Q(\sqrt{-1}\,\lambda)$ can be represented by a Jacobi field, and combining with (2.27), we see that the finite dimensional part of the Jacobian of the transformation of variables stated in Subsecton 2.4 can be constructed by using Jacobi fields.

In this case, it follows from the form of $L(x, \dot{x})$ in (2.13) that the Euler-Lagrange equation is

$$\left\{\left(\frac{d}{ds}\right)^2 - \sqrt{-1}\,\lambda\left(\frac{d}{ds}\right)J\right\}{}^t u = 0, \tag{2.28}$$

and the Jacobi equation is of the same form.

Let $Y(s,p)$ be the solution of the Jacobi equation satisfying the condition $Y(0,p)$
$= 0$, $\dot{Y}(0,p) = p \in \mathbb{R}^2$, and define an operator $J(\sqrt{-1}\,\lambda)$ by $J(\sqrt{-1}\,\lambda)[{}^t p] = {}^t Y(1,p)$, $p \in \mathbb{R}^2$. We define an operator $\mathcal{L}$ by

$$\mathcal{L}[h] = {}^t\bigl(((d/ds)^2 - \sqrt{-1}\,\lambda(d/ds)J)[{}^t h]\bigr).$$

Then

$$\begin{aligned} &\mathcal{L} = \frac{d^2}{ds^2}(I - 2\sqrt{-1}\,\lambda B^{\#}), \\ &\ker(\mathcal{L}|_H) = (I - 2\sqrt{-1}\,\lambda B^{\#})^{-1}((H^0)^{\perp}) = (I - 2\sqrt{-1}\,\lambda B_0)^{-1}((H^0)^{\perp}). \end{aligned} \tag{2.29}$$

It follows that

$$J(\sqrt{-1}\,\lambda) = {}^t Q(\sqrt{-1}\,\lambda) \qquad (= \text{the transposed matrix of } Q(\sqrt{-1}\,\lambda)). \tag{2.30}$$

Calculating the Van Vleck matrix using (2.10), by (2.26) and (2.30), we can show that

$$J(\sqrt{-1}\,\lambda) = \left(\frac{\partial S_{\mathrm{cl}}}{\partial x \partial y}\right)^{-1}. \tag{2.31}$$

To accomplish our objective, we must prove (2.31) directly, using the definition of Jacobi fields without (2.10). This is possible if we use a relation analogous to the relation called in [**43**] the second variational formula (see [**22**] for details). Starting from this relation and reversing the above line of reasoning, we can arrive at the Van Vleck formula by combining (2.25) and (2.30).

REMARK 2.2. The relation between the Van Vleck matrix and Jacobi fields for the propagator was discussed in C. De Witt-Morette [**10**], etc. We also mention that in B. De Witt [**9**] the Van Vleck formula was discussed using the decomposition of the integral kernel corresponding to the Jacobi equation (2.28) instead of the Hilbert-Schmidt operator introduced in 2.4. This method is related to the idea described above through (2.29).

2.7. Cameron-Martin transformation. For the harmonic oscillator, a method that shows more clearly that the determinant in (2.27) is a Jacobian determinant of the transformation of variables is known ([**4**]). This corresponds to a shift method in path integrals (cf. [**5**]). The same method applies to the case of uniform magnetic fields. For simplicity, we treat the case of an integral of Laplace type, replacing $\sqrt{-1}\,\lambda$ by a real number z in the integral of the left-hand side of (2.4). According to this, we change $\sqrt{-1}\,\lambda$ to a real number z in (2.13) and consider the L thus obtained. Then we obtain the Jacobi equation as in the preceding subsection:

$$\left\{\Big(\frac{d}{ds}\Big)^2 - z\Big(\frac{d}{ds}\Big)J\right\}{}^t u = 0.$$

We consider $z \in \mathbb{R}$ whose absolute value $|z|$ is sufficiently small. Under this assumption, using two independent solutions $Y_1(s) = (1 + \cos zs, -\sin zs)$, $Y_2(s) = (\sin zs, 1 + \cos zs)$ of the above equation, we define transformations $\mathcal{T} : W \to W$ and $\mathcal{K} : W \to W$ as follows:

$$\mathcal{T}[w](s) = -\int_0^s {}^t\left\{\left[\frac{\partial}{\partial u}A(u)^{-1}\right]({}^t w(u))\right\}du,$$
$$\mathcal{K}[w](s) = -\int_0^s {}^t\{K(u)({}^t w(u))\}du, \qquad w \in W.$$

Here ${}^tA(s) = \big({}^tY_1(s), {}^tY_2(s)\big)$ and

$$K(u) = -\Big(\frac{\partial}{\partial u}A(u)\Big)A(u)^{-1} = \frac{z}{2}\begin{pmatrix} \tan(uz/2) & 1 \\ -1 & \tan(uz/2)\end{pmatrix}.$$

If we set $\mathcal{M} = I + \mathcal{K}$ and $W^0 = \{w; w \in W, w(1) = 0\}$, then we have

$$\langle (I - 2\sqrt{-1}\,\lambda B^{\#})[h], h\rangle = \langle \mathcal{M}[h], \mathcal{M}[h]\rangle, \quad h \in H^0$$

(diagonalization of a quadratic form), and

$$\mathcal{M}[(I + \mathcal{T})[w]] = w, \quad w \in W^0.$$

By calculating the determinant of $\mathcal{M}$ and the variance of $(I + \mathcal{T})[w](1)$, from the above relations we obtain

$$\int_W \exp\big[zA(1; w)\big]\delta_0(w(1))P(dw)$$
$$= \cos\Big(\frac{z}{2}\Big)\int_W \delta_0((I + \mathcal{T})[w](1))P(dw) = \frac{1}{2\pi}\frac{z/2}{\sin(z/2)}.$$

By a close examination of the above reasoning, we find that this equality holds for $|z| < 2\pi$. This equality is generally called Lévy's formula, including this case.

3. General framework

In this section, we give a general framework for our problem according to [**19, 20, 22**]. As in Section 2, $W = \{w; w \in C([0,t] \to \mathbb{R}^2), w(0) = 0\}$. Let (W, P) be the Wiener space and H the Cameron-Martin subspace of W. We use the same notation H for the complexification of H, as stated in 2.4. Let $L^2(W, P)$ be the Hilbert space of square integrable (with respect to P) functions on W, and denote its Wiener-Itô decomposition in terms of Hermite polynomials by

$$L^2(W,P) = \sum_{n=0}^{\infty} \oplus \mathcal{C}_n. \tag{3.1}$$

We refer to, for example, [**24**] for the definition. It is known that for any quadratic Wiener functional $F(w) \in \mathcal{C}_2$, there exists a $B \in \mathrm{HS}^{(s)}(H)$ such that

$$\begin{aligned} F(w) = \sum_{j,k=1}^{2} \Big[\int_0^t \int_0^s & F^{jk}(s,u) dw^k(u) dw^j(s) \\ & + \int_0^t \int_0^u F^{jk}(s,u) dw^j(s) dw^k(u) \Big], \\ \frac{d}{ds} B[h]^{(j)}(s) = \sum_{k=1}^{2} \int_0^t & F^{jk}(s,u) \dot{h}^k(u) du, \qquad h \in H \end{aligned} \tag{3.2}$$

(see for example [**19, 20, 22**]). For simplicity, we use the notation $F(w) = \langle Bw, w \rangle$. Let $\eta = \{h_1, h_2, \ldots, h_m\}$ be a linearly independent system in H, and set $\eta(w) = (\langle h_1, w\rangle, \langle h_2, w\rangle, \ldots, \langle h_m, w\rangle)$. Let $\eta\eta^*$ be a symmetric matrix whose (i,j) element, $i, j = 1, 2, \ldots, m$, is given by $\langle h_i, h_j\rangle$. We denote by $H(\eta)$ the closed subspace of H generated by η, and by $H(\eta)^\perp$ the orthogonal complement of $H(\eta)$ in H, and finally set $B^\# = P_{H(\eta)^\perp} B P_{H(\eta)^\perp}$, $B_\eta = P_{H(\eta)} B P_{H(\eta)}$. Let $\{\lambda_n\}$ be the eigenvalues of $B^\#$, and set $\sigma(B^\#) = \max\{|\lambda_n|\}$. In this notation, using the Fourier expansion of Brownian motion by means of an orthonormal basis consisting of the system of eigenfunctions of $B^\#$ and the system of orthonormalized η, according to the same idea as in 2.4, we can obtain the following result ([**19, 20, 22, 29, 30**]).

THEOREM 3.1 ([**20, 22**]). *For $z \in \mathbb{C}$ with $2|\mathrm{Re}(z)|\sigma(B^\#) < 1$, we have*

$$\begin{aligned} & \int_W \exp\big[z\langle Bw, w\rangle\big] \delta_0(\eta(w)) P(dw) \\ & = \Big(\frac{1}{2\pi}\Big)^{m/2} \big\{\det(\eta\eta^*) \det{}_2(I - 2zB^\#)\big\}^{-1/2} \exp[-z\mathrm{tr}(B_\eta)]. \end{aligned} \tag{3.3}$$

For the proof, see [**20, 22**]. It is necesarry to modify [**20**] according to [**22**].

Now, in order to make the problem tractable, we make an assumption introduced in [**19**], as the decomposition (2.18) suggests.

ASSUMPTION 3.1. $B \in \mathrm{HS}^{(s)}(H)$ can be decomposed as

$$B = B_0 + B_1, \tag{3.4}$$

where B_0 is a Volterra operator, and the dimension of the range of B_1 is finite.

In the following, we always take η to be linearly independent and to generate the range H_1 of B_1. That is, we take η such that $H(\eta) = H_1$. We then set

$$B^{\#} = B_0 + B_1^{\#}, \quad B_1^{\#} = -P_{H(\eta)}B_0. \tag{3.5}$$

We can prove the following theorem using the method developed in [**19, 20, 22**].

THEOREM 3.2 ([**20, 22**]). *Let Assumption 3.1 be satisfied. For $z \in \mathbb{C}$ satisfying $2|\mathrm{Re}(z)|\sigma(B^{\#}) < 1$, we have*

$$\begin{aligned} &\int_W \exp[z\langle Bw, w\rangle]\delta_0(\eta(w))P(dw) \\ &\quad = \left(\frac{1}{2\pi}\right)^{m/2} \{\det(\eta\eta^*)\det Q(z)\}^{-1/2} \exp[-z\mathrm{tr}(P_{H(\eta)}B_1)], \end{aligned} \tag{3.6}$$

where $Q(z) = I - 2zB_1^{\#}(I - 2zB_0)^{-1}$.

The theorem is a consequence of Theorem 3.1 via a formal operator calculus, if we note that the same relation $I - 2zB^{\#} = Q(z)(I - 2zB_0)$ as (2.24) holds and B_0 is a Volterra operator (for details, see[**20, 22**]).

In the argument of this paper, the equality (3.6) plays a central role.

4. Case of a quadratic Lagrangian

In electromagnetic fields, a concrete expression of the propagator has been calculated not only in the cases of uniform magnetic field and harmonic oscillator, but in the cases when the Lagrangians are linear combinations of them. On the other hand, in the case of the Wiener space, Fourier transforms of probability distributions of typical quadratic Wiener functionals began to be studied by Cameron and Martin and others in the mid-40s, and the study was continued by M. Kac, P. Lévy and many others ([**28, 34**], etc). Further, from the mid-80s, related with the study of polymers, the asymptotic behaviour of eigenvalues of the Schrödinger operator with electromagnetic fields, and stochastic oscillatory integrals, these studies started to attract many people (see [**1, 6, 7, 13, 37, 56**] and their references). These studies, following the idea stated in Section 2, can be treated in a unified way within the framework given in Section 3. Let us consider

$$\begin{aligned} F_t(w_x; \mu, \gamma) &= \mu A(t; w_x) - \sum_{k=1}^{2} \gamma_k \int_0^t (w_x^k(s))^2 ds \quad (\in \mathcal{C}_0 \oplus \mathcal{C}_1 \oplus \mathcal{C}_2), \\ F_t^*(w; \mu, \gamma) &= \mu A(t; w) - \sum_{k=1}^{2} \gamma_k \int_0^t (w^k(s))^2 ds + \frac{t^2(\gamma_1 + \gamma_2)}{2} \quad \in \mathcal{C}_2. \end{aligned} \tag{4.1}$$

The operator $B \in \mathrm{HS}^{(\mathrm{s})}(H)$ corresponding to $F_t^*(w; \mu, \gamma)$ is given by the following equation:

$$\begin{aligned} B &= B_0 + B_1, \\ B_0[h] &= \frac{\mu}{2}\mathcal{I}[{}^t(J[{}^t h])] + \mathcal{I}^2[{}^t(G[{}^t h])], \qquad G = \begin{pmatrix} \gamma_1 & 0 \\ 0 & \gamma_2 \end{pmatrix}, \\ B_1[h](s) &= -\frac{\mu}{4}\,{}^t(J[{}^t h])(t)s - \mathcal{I}[{}^t(G[{}^t h])](t)s, \end{aligned}$$

where $\mathcal{I}$ was introduced in 2.4. Using $\phi_{0,i}$, $i = 1, 2$, which are given in (2.21), we set $\eta = \{h_1, h_2\}$, $h_i = \phi_{0,i}$, $i = 1, 2$. Then $H(\eta) = (H^0)^{\perp}$. We consider $B_1^{\#}$, $Q(z)$, etc., determined by (3.5), (3.6). We define L corresponding to $F_t(w_x;\mu,\gamma)$ in (4.1) as

$$L(x,\dot{x}) = \frac{1}{2}\big((\dot{x}^1)^2+(\dot{x}^2)^2\big) - z\Big[\frac{\mu}{2}(x^1\dot{x}^2 - x^2\dot{x}^1) - \big(\gamma_1(x^1)^2+\gamma_2(x^2)^2\big)\Big], \quad x=(x^1,x^2),\ \dot{x}=(\dot{x}^1,\dot{x}^2), \tag{4.2}$$

then define $S(t,x,y;\phi)$ on $\Omega^*(t,x,y)$ by (2.8), and denote the critical point by $\phi_{\rm cl}$.

With this notation, as in Section 2 we see that the Euler-Lagrange equation and the Jacobi equation are both of the form

$$\Big\{\Big(\frac{d}{ds}\Big)^2 - z\mu\Big(\frac{d}{ds}\Big)J - 2zG\Big\}{}^t u = 0. \tag{4.3}$$

The Jacobi field obtained by the variation of $\phi_{\rm cl}$ is a solution of this equation. We denote by $Y(s,p)$ a solution of the Jacobi equation satisfying $Y(0,p)=0$, $\dot{Y}(0,p)=p\in\mathbb{R}^2$, and we define a mapping $J_t(z)$ by $J_t(z)[{}^tp] = {}^tY(t,p)$, $p\in\mathbb{R}^2$. Next, we take two points $a=\phi(u)$, $b=\phi(v)$, $u\neq v$, on a classical trajectory ϕ that is a solution of (4.3). We call a and b conjugate along ϕ if there exists a nonvanishing solution of the Jacobi equation that vanishes at $s=u$ and $s=v$ along ϕ. Let $\sigma^*(z) = \inf\{s; x=\phi(0)$ and $y=\phi(s)$ are conjugate along a classical trajectory $\phi\}$, and set $S_{\rm cl}(t,x,y) = S(t,x,y;\phi_{\rm cl})$.

THEOREM 4.1 (Van Vleck formula [**20, 22**]). *Let $0<t<\sigma^*(z)$. Then we have*

$$\int_W \exp\big[zF_t(w_x;\mu,\gamma)\big]\delta_y(w_x(t))P(dw) = \frac{1}{2\pi}\Big\{\det\Big(\frac{\partial^2 S_{\rm cl}}{\partial x\partial y}\Big)\Big\}^{1/2}\exp\big[-S_{\rm cl}(t,x,y)\big], \tag{4.4}$$

$$J_t(z) = t\,{}^tQ(z), \quad J_t(z) = \Big(\frac{\partial^2 S_{\rm cl}}{\partial x\partial y}\Big)^{-1}. \tag{4.5}$$

Idea of proof. For the details of the proof, see [**20, 22**]. The ideas stated in each subsection of section 2 also work well in this case. There are several methods that can be used for the proof. If we follow any one of them, the starting point, which is the Cameron-Martin idea, is to establish the equality

$$\int_W \exp\big[zF_t(w_x;\mu,\gamma)\big]\delta_y(w_x(t))P(dw) = \int_W \exp\big[zF_t(w;\mu,\gamma)\big]\delta_0(w(t))P(dw)\,\exp[-S_{\rm cl}(t,x,y)]. \tag{4.6}$$

The method of proving the Van Vleck formula for the propagator works in this case (see [**50**]). Take a z with sufficiently small $|z|$ and set

$$A(t) = \int_W \exp\big[zF_t(w;\mu,\gamma)\big]\delta_0(w(t))P(dw).$$

Then, combining (4.6) and the heat equation determined by L in (4.2), we have

$$\frac{1}{A(t)}\frac{\partial A(t)}{\partial t} = -\frac{1}{2}\Delta S_{\rm cl}(t,x,y).$$

Using this equation and the symmetry of the solution of the heat equation with respect to x and y, we obtain (4.4) (see [**20, 49**]). The proof of this method does not make explicit the above mentioned meaning of the Van Vleck formula. So, following the idea stated just before Remark 2.2 in Subsection 2.6, we want to prove (4.5) first, and then obtain (4.4) by combining with (3.6). After the case in Subsection 2.6, we define an operator $\mathcal{L}$ corresponding to (4.3) by

$$\mathcal{L}[h] = {}^t\big(((d/ds)^2 - z\mu(d/ds)J - 2zG)[{}^t h]\big).$$

Then the same equation as (2.29) holds. We can see that

$$\widetilde{Y}(s;\phi^*) = (I - 2zB^\#)^{-1}[\phi^*],$$

$\phi^* = Q(z)\big(\sum_{i=1}^2 p_i h_i\big)$, $p = (p_1, p_2) \in \mathbb{R}^2$, is a Jacobi field, which we used to define the mapping $J_t(z)$. It follows that the first equation of (4.5) holds. The second equation of (4.5) can be derived by using the second variational formula stated in the proof of (2.31) (for details, see [**22**]). Also, noting that

$$\operatorname{tr}\big(P_{H(\eta)} B_1 P_{H(\eta)}\big) = -\frac{t^2(\gamma_1 + \gamma_2)}{2}$$

by (3.6), we can show that

$$\int_W \exp\big[zF_t(w;\mu,\gamma)\big]\delta_0(w(t))P(dw) = \frac{1}{2\pi t}\{\det Q(z)\}^{-1/2}.$$

Summarizing these results, we obtain (4.4).

REMARK 4.1. Equation (4.5) gives a method that enables us to obtain the Van Vleck matrix directly from the symmetric Hilbert-Schmidt operator, which is a characteristic of the quadratic Wiener functional. We mention that the above discussion implies

$$\det{}_2(I - 2zB^\#) = \det\big((I - 2zB^\#)|_{H^*}\big), \qquad H^* = \ker\big(\mathcal{L}|_H\big). \tag{4.7}$$

REMARK 4.2. As already mentioned, the identity corresponding to (4.4) is well known for path integrals. Substantially the same facts are often used in the asymptotic theory. We refer to Kac [**29, 30**], Molchanov [**44**], Rogers and Shi [**48**], etc.

EXAMPLE 4.1. The quantities in the right-hand side of (4.4) and in (4.5) can be concretely calculated by using (4.2) and (4.3), so we can obtain explicit formulas for known examples as special cases by (4.4) and (4.5). For example, besides the typical example stated in Section 2, the most famous one is the case of the harmonic oscillator.

If we set $F_t(w_x) = F_t(w_x; 0, (1,1))$, then for $z \in \mathbb{R}$, $-2zt^2\pi^{-2} < 1$, $t > 0$, $x, y \in \mathbb{R}^2$, we have

$$\begin{aligned}\int_W &\exp\big[zF_t(w_x)\big]\delta_y(w_x(t))P(dw)\\ &= \frac{1}{2\pi t}\frac{\sqrt{2z}\,t}{\sinh(\sqrt{2z}\,t)}\\ &\quad\times\exp\left[-\sqrt{2z}\,t\coth(\sqrt{2z}\,t)\,\frac{|x|^2 - 2(x,y)\operatorname{sech}(\sqrt{2z}\,t) + |y|^2}{2t}\right]\end{aligned} \tag{4.8}$$

(for example, see [**20**]). We mention the following as another example: Let $z = \sqrt{-1}\,\lambda$, $\mu = 1$, $\gamma_1 = -\gamma_2 = \gamma/(2\lambda)$, $\lambda, \gamma > 0$, $\lambda^2 > 2\gamma$. Then we can explicitly calculate the right-hand side of (4.4) ([**20**]). The path integral counterpart for this has been calculated in connection with the study of polymers (for example, we refer the reader to [**32**]). Computations of the path integrals for a more general quadratic Lagrangian are collected in Chapter 6, Section 6.2 of [**15**], and for Wiener integrals, see for example [**1, 2, 6, 20, 48, 56**].

REMARK 4.3. We make a remark on the validity of (4.4). Since $L(x, \dot{x})$ in (4.2) contains a parameter z, $\sigma^*(z)$ depends on z, and hence the extent of t for which (4.4) holds depends on z. By the definition of (4.6) and $B^\#$, (4.4) holds if the inequality $2|\mathrm{Re}(z)|\sigma(B^\#) < 1$ is satisfied. Therefore it holds for z purely imaginary, $z = \sqrt{-1}\,\lambda$. In the case of $z \in \mathbb{R}$, if $\mu = 1$ and $\gamma_1 = \gamma_2 = 0$, then, since $2|\mathrm{Re}(z)|\sigma(B^\#) = t|z|/(2\pi)$, the condition $2|\mathrm{Re}(z)|\sigma(B^\#) < 1$ becomes $0 < t|z| < 2\pi$ (see Subsection 2.7).

In the case when $z \in \mathbb{R}$, $\mu = 0$, $\gamma_1 = \gamma_2 = 1$, as stated in Example 4.1, if $-2zt^2 < \pi^2$, then we have (4.4). Thus (4.8) holds. If these conditions are not satisfied, the integral of the right-hand side of (4.4) may diverge. For examlple, see [**20, 25**].

As for the problem of validity beyond the conjugate point, the situations are entirely different between Wiener integrals and path integrals. The latter case involves some interesting phenomena (for these topics, see for example [**29, 30, 47**]).

These problems are deeeply related to the behaviour of the classical trajectory. We have no room here to discuss this, but for example in the case $\mu = 1$, $\gamma_1 = \gamma_2 = 0$, $z > 0$, as is well known in the case of a uniform magnetic field, the classical trajectory ϕ satisfying $\phi(0) = x$, $x = (x^1, x^2) \in \mathbb{R}^2$, $\dot{\phi}(0) = p$, $p = (p^1, p^2) \in \mathbb{R}^2$, is a circle with center $(x^1 - p^2/2, x^2 + p^1/2)$ and radius $|p|/z$. This fact is related to the observation in the following Remark 4.4:

REMARK 4.4. The quadratic Wiener functionals treated in this section were studied by Matsumoto and Ueki [**41**] from another point of view.

Also, for harmonic oscillators and uniform magnetic fields, Hermite polynomials and complex Hermite polynomials appear as the corresponding eigenfunctions. As a generalization of this fact, Matsumoto [**40**] studied the Lagrangian of (4.2).

REMARK 4.5. In this section we have dealt with only some typical examples of Wiener functionals corresponding to a quadratic Lagrangian. We can also treat systematically, in the framework stated in Section 3, many examples containing those discussed by many authors [**1, 6, 11, 48, 56**] (for the details, see [**20**]). Among them an expression called the generalized Lévy formula (see [**1, 11, 20, 56**]) is included.

5. Case of Gaussian process

What we discussed in the previous section remains valid for general quadratic Lagrangians. Hence it is described in terms of the theory of partial differential equations of second order. But in the case of Feynman path integrals, one deals with nonlocal potentials and so on, which exceed the above framework (see for example, [**32**]). Even so, our setting in Secton 3 can treat such topics, with some limitations.

For example, the following case, treated in [**19, 20**], is a typical example.

In this section, for simplicity, we restrict ourselves to the case of $t = 1$. That is, we consider $W = \{w; w \in C([0,1] \to \mathbb{R}^2), w(0) = 0\}$. Let N be a nonnegative integer.

As in subsection 2.4, we define a mapping $\mathcal{I} : W \to W$ by

$$\mathcal{I}[w](t) = \int_0^t w(s)ds, \quad 0 \leqq t \leqq 1,$$
$$\mathcal{I}^n = \mathcal{I}\mathcal{I}^{n-1}, \quad 1 \leqq n \leqq N, \quad \mathcal{I}^0[w] = w, \quad w \in W.$$

Then $X_N(t,w) = \mathcal{I}^N[w](t)$, $0 \leqq t \leqq 1$, is a two-dimensional Gaussian process with zero mean vector and covariance matrix

$$\begin{pmatrix} \Gamma(t,s) & 0 \\ 0 & \Gamma(t,s) \end{pmatrix},$$
$$\Gamma(t,s) = \frac{1}{(N!)^2}\int_0^{t\wedge s}(t-u)^N(s-u)^N du, \quad 0 \leqq t,s \leqq 1.$$

In this case, if we denote by $A_N(w)$ the stochastic line integral of a differential one-form $\theta = (1/2)(x^1dx^2 - x^2dx^1)$, $x = (x^1,x^2) \in \mathbb{R}^2$, on $\mathbb{R}^2$ along a curve $X_N[0,1] = \{X_N(t,w); 0 \leqq t \leqq 1\}$, we see that the operator $B \in \mathrm{HS}^{(\mathrm{s})}(H)$ for

$$F_N(w) = (-1)^N A_N(w) \in \mathcal{C}_2$$

given in Section 3 is of the following form:

$$\begin{aligned} &B = B_0 + B_1, \\ &\frac{d}{dt}B_0[h](t) = \frac{1}{2}\mathcal{I}^{2N}\left[{}^t(J[{}^th])\right](t), \\ &\frac{d}{dt}B_1[h](t) = -\int_0^1 a(1-t,1-s){}^t(J[{}^th])(s)ds, \\ &a(t,s) = \frac{1}{2(2N)!}\left[\sum_{k=1}^{N-1}\binom{2N}{k}s^{2N-k}(-t)^k + \frac{1}{2}\binom{2N}{N}s^N(-t)^N\right], \\ &\qquad\qquad 0 \leqq t,s \leqq 1. \end{aligned} \tag{5.1}$$

Consequently, Assumption 3.1 is satisfied. So the framework in Section 3 works. Set $\phi_k(t) = t^k$, $0 \leqq t \leqq 1$, $k = 0,1,\dots,N$, $\eta = \{h_0,h_1,\dots,h_{2N+1}\}$, where

$$h_{2k} = \mathcal{I}[\phi_k](1,0), \quad h_{2k+1} = \mathcal{I}[\phi_k](0,1), \qquad k = 0,1,\dots,N.$$

Then $H(\eta)$ coincides with the range H_1 of B_1. We define $B^\#$, $B_1^\#$ and $Q(z)$ as in Section 3. We define a quadratic form $S_N(h;z,B)$, $z \in \mathbb{C}$, corresponding to $F_N(w)$ by

$$S_N(h;z,B) = \frac{1}{2}\langle (I - 2zB)h, h\rangle, \quad h \in H.$$

Then its Euler-Lagrange equation is

$$\left\{\left(\frac{d}{ds}\right)^{2(N+1)} - zJ\left(\frac{d}{ds}\right)\right\}{}^tu(s) = 0, \qquad 0 \leqq s \leqq 1. \tag{5.2}$$

The Jacobi equation has the same form. Now let

$$\eta(h) = \big(\langle h_0,h\rangle, \langle h_1,h\rangle, \dots, \langle h_{2N+1},h\rangle\big).$$

For $x \in \mathbb{R}^{2(N+1)}$, if there exists a critical point ϕ_{cl} of $S_N(h;z,B)$ belonging to $H(\eta,x) = \{h \in H; \eta(h) = x\}$ (that is, if there exists a solution of (5.2)), then we set $S_{\mathrm{cl}}(x;z,B) = S_N(\phi_{\mathrm{cl}};z,B)$. For $p = (p_0,p_1,\dots,p_{2N+1}) \in \mathbb{R}^{2(N+1)}$, we denote by $Y(s,p)$, $0 \leqq s \leqq 1$, the Jacobi field obtained as a solution of (5.2) satisfying the following condition:

$$
\begin{aligned}
&Y(0,p) = 0, \\
&\Big(\Big(\frac{d}{ds}\Big)^k Y\Big)(0,p) = k!(p_{2(k-1)}, p_{2(k-1)+1}), \quad k = 1,2,\ldots,N+1, \\
&\Big(\Big(\frac{d}{ds}\Big)^k Y\Big)(0,p) = 0, \quad k = N+2,\ldots,2N+1,
\end{aligned} \tag{5.3}
$$

and we define a mapping $J(z)$ by

$$J(z)[{}^t p] = {}^t(\langle Y(\cdot,p), h_0\rangle, \langle Y(\cdot,p), h_1\rangle, \ldots, \langle Y(\cdot,p), h_{2N+1}\rangle), \quad p \in \mathbb{R}^{2(N+1)}.$$

We can prove the following.

THEOREM 5.1 (Van Vleck formula [**19, 22**]). *Let $z \in \mathbb{C}$ satisfy the condition*

$$2|\mathrm{Re}(z)|\sigma(B^{\#}) < 1.$$

Then

$$
\begin{aligned}
&\int_W \exp[zF_N(w)]\delta_x(\eta(w))P(dw) \\
&\qquad = \Big(\frac{1}{2\pi}\Big)^{N+1} \{\det J(z)\}^{-1/2} \exp[-S_{\mathrm{cl}}(x;z,B)], \\
&J(z) = A^t Q(z), \qquad A = (\langle h_i, h_j\rangle)_{0 \leq i,j \leq 2N+1}.
\end{aligned} \tag{5.4}
$$

The proof is carried out like that of Theorem 4.1: we follow the line stated in Subsection 2.6. In fact, if we define the operator $\mathcal{L}$ corresponding to (5.2) by

$$\mathcal{L}[h] = {}^t\big(\big((d/ds)^{2(N+1)} - zJ(d/ds)\big)[{}^t h]\big),$$

then the equality $\mathcal{L} = (d/ds)^{2(N+1)}(I - 2zB^{\#})$ and the counterpart of the second equality in (2.29) hold. Now for $\phi^* = Q(z)\big[\sum_{i=0}^{2N+1} p_i h_i\big]$, $p = (p_0, \ldots, p_{2N+1}) \in \mathbb{R}^{2(N+1)}$, setting $\widetilde{Y}(s;\phi^*) = (I - 2zB^{\#})^{-1}[\phi^*]$, we can see that $\widetilde{Y}$ is the Jacobi field satisfying (5.3).

Combining this and Theorem 3.2, we can complete the proof.

Moreover, using a similar idea as in Subsections 2.4, 2.5, and Section 3, we can show the following:

$$
\begin{aligned}
&\int_W \exp[\sqrt{-1}\,\lambda F_N(w)]P(dw) \\
&\qquad = \int_{\mathbb{R}^{2(N+1)}} \int_W \exp[\sqrt{-1}\,\lambda F_N(w)]\delta_x(\eta(w))P(dw)dx \\
&\qquad = \big\{\det{}_2(I - 2\sqrt{-1}\,\lambda B)\big\}^{-1/2} = \big\{\det Q^*(\sqrt{-1}\,\lambda)\big\}^{-1/2}, \\
&Q^*(\sqrt{-1}\,\lambda) = I - 2\sqrt{-1}\,\lambda B_1(I - 2\sqrt{-1}\,\lambda B_0)^{-1}.
\end{aligned} \tag{5.5}
$$

Furthermore, in Ikeda-Kusuoka-Manabe [**19**] the following expression of the finite dimensional determinant $\det Q^*(\sqrt{-1}\,\lambda)$ and a way to compute the quantities

$c(A)$, $\tau(A)$, $\sigma(A)$, Λ_0 and Λ_1 in the expression are given:

$$\det Q^*(\sqrt{-1}\,\lambda) = (-1)^{N(N-1)/2}\gamma(N)^N \Big[\sum_{A\in\Lambda_0} c(A)\cosh\big(\tau(A)\lambda^{\gamma(N)}\big)$$
$$+ 2\sum_{A\in\Lambda_1} c(A)\cosh\big(\tau(A)\lambda^{\gamma(N)}\big)\cos\big(\sigma(A)\lambda^{\gamma(N)}\big)\Big], \tag{5.6}$$
$$\gamma(N) = \frac{1}{2N+1}.$$

Let us consider a configuration of $(N+1)$ balls placed on lattice points on the circle with center at the origin and radius $(2N+1)/(2\pi)$. At each time, we move a ball one step without passing over any other balls. Then the total number of possible moves of balls depends on N, and there exists a matrix determined by the number of moves from one configuration to another. We can then express the coefficients in (5.6) more explicitly with the help of the eigenvalues of this matrix. The matrix has an extreme characteristic, and it is known that if N is small, then its eigenvalues and eigenvectors can be calculated. But, in the general case, these questions remain open. For details, see [**19**].

REMARK 5.1. It is well known in the theory of the stationary phase method that in the finite dimensional case, the asymptotic behavior of oscillatory integrals with nondegenerate phase functions depends on the dimension (see [**18**] for examples). Concerning the left-hand side of (5.5), which is considered as an example of an infinite dimensional oscillatory integral, we can see from (5.6) that its speed of decrease as λ tends to infinity depends on N, that is, smoothness of the sample function of the Gaussian process (for details, see [**21**]).

6. Selberg trace formula

In the theory of path integrals, the Van Vleck formula is used to show that the quantum state is determined by the corresponding classical one if the Lagrangian is quadratic, as stated. In this case, the fact that the problem is reduced to the calculation of integrals of Gaussian type plays an essential role. We cannot expect that the same situation occurs in the case of general Lagrangians, which may not be quadratic, nor in the case of a curved Riemannian metric. So in the semiclassical approximation, which studies what happens when the Planck constant $\hbar$ tends to zero, one analyses the asymptotics using the *Van Vleck* propagator (or Van Vleck approximation) in case of Riemannian space (N, g),

$$K^{\#}(t,x,y) = \frac{1}{\hbar 2\pi\sqrt{-1}} \left\{ \left(\frac{\partial^2 S_{\mathrm{cl}}}{\partial x \partial y} \right) \right\}^{1/2}$$
$$\times \exp\Big[-\frac{\sqrt{-1}}{\hbar} S_{\mathrm{cl}}(t,x,y) \Big] (G(x)G(y))^{-1/4},$$

instead of the propagator ([**50**]). Here $G(x) = \det g(x)$, $g(x) = (g_{ij}(x))$, and $S_{\mathrm{cl}}(t,x,y)$ means the value of the action integral corresponding to the (general) Lagrangian determined by g at the classical trajectory. $K^{\#}(t,x,y)$ is determined by the quantities of classical dynamics. Moreover, in general, it is different from the propagator $K(t,x,y)$. For the method using the Van Vleck propagator to be effective, roughly speaking, it is necessary that an analogue to the stationary phase method in finite dimensional spaces works for the path integral. But it is not easy to establish this in the path space. It is known that several of the characteristics determined by the propagator possess a classical dynamical character. For example, the function $x/\sin x$, which appears in the Van Vleck determinant in the case of a uniform magnetic field, plays an essential role in the main term of the asymptotics ([**24**]). Furthermore, this situation is clearer, as M. C. Gutzwiller et al. have discussed in the study of Selberg's trace formula. We mention that the role of the Jacobi field in the theory of asymptotics of heat equations on Riemannian spaces is systematically discussed in a research paper by S. A. Molchanov in connection with the result of S. Minakshisundaram and A. Pleijel ([**44**]).

Let us consider the Laplace-Beltrami operator Δ detemined by the Poincaré metric on the Poincaré upper half plane $\mathbb{H}^2$. We denote by $p(t, z_1, z_2)$ the fundamental solution (with respect to the Riemannian volume) of the corresponding heat equation. As a Riemannian manifold, we consider a compact Riemann surface of genus greater than 2 realized as a quotient space of $\mathbb{H}^2$ by a discrete subgroup Γ. Let $\{\lambda_n\}$ be the eigenvalues of $\Delta/2$ on M; then A. Selberg [**51**] has shown the following equality.

Selberg's trace formula ([51]). For any $t > 0$,

$$
\begin{aligned}
&\sum_{k=1}^{\infty} \exp[-t\lambda_n] \\
&\qquad = \frac{\mathrm{vol}(M)}{2\pi} \exp\Big[-\frac{t}{8}\Big] \int_0^\infty \frac{x}{\cosh x} \exp\Big[-\frac{x^2}{2t}\Big] \sinh 2\pi x \, dx \\
&\qquad + \sum_{n=1}^{\infty} \sum_{\gamma \in \Gamma_0} \ell(\gamma) I_1(t, n\ell(\gamma)) I_2(t, n\ell(\gamma)),
\end{aligned} \tag{6.1}
$$

$$
\begin{aligned}
\ell(\gamma) &= \inf\{d(z, \gamma z); z \in \mathbb{H}^2\}, \\
I_1(t, x) &= \Big\{\frac{x}{\sinh x}\Big\}^{1/2} \frac{1}{2\pi t} \exp\Big[-\frac{t}{8} - \frac{x^2}{2t}\Big], \\
I_2(t, x) &= (2\pi t)^{1/2} \Big[\frac{\sinh x}{(4x(\sinh x/2))^2}\Big]^{1/2}.
\end{aligned}
$$

Here, Γ_0 is a set of primitive elements of Γ that are not conjugate to each other, and d is the metric function on $\mathbb{H}^2$. Gutzwiller et al. pointed out that the trace formula (considered as semiclassical approximation) obtained by the Van Vleck approximation contains a term of the same form as the second term in the right-hand side of the first equation (cf. [**8, 16, 17**]). As is well known, the Van Vleck approximation of $p(t, z_1, z_2)$ is

$$
p^{\#}(t, z_1, z_2) = I_1(t, d(z_1, z_2)), \quad z_1, z_2 \in \mathbb{H}^2.
$$

On the other hand, by definition, for given $\gamma \in \Gamma_0$ there exists an $a = a(\gamma) > 1$ such that γ is conjugate with the 2×2 diagonal matrix whose elements are given by $a(\gamma)$, $1/a(\gamma)$ and $\ell(\gamma) = \log a(\gamma)^2$. Furthermore, by using the fact that the fundamental solution of the heat equation on M can be constructed from $p(t, z_1, z_2)$, we see that, as McKean [**42**] has shown, the second term in the right-hand side in (6.1) is a modification of

$$\sum_{n=1}^{\infty}\sum_{\gamma\in\Gamma_0}\int_1^{a^2}\frac{1}{y^2}dy\int_{\mathbb{R}}p(t,x+\sqrt{-1}\,y,a^{2n}(x+\sqrt{-1}\,y))dx.$$

If we replace p by $p^{\#}$, we have

$$\begin{aligned}&\int_{\mathbb{R}}p^{\#}(t,x+\sqrt{-1}\,y,a^{2n}(x+\sqrt{-1}\,y))dx\\&\quad=yI_1(t,n\ell(\gamma))\int_{\mathbb{R}}\left\{\frac{d(x)}{\sinh d(x)}\,\frac{\sinh d(0)}{d(0)}\right\}^{1/2}\exp\left[-\frac{(d(x)^2-d(0)^2)}{2t}\right]dx,\\&\qquad\qquad d(x)=d(x+\sqrt{-1}\,,a^{2n}(x+\sqrt{-1}\,)).\end{aligned}$$

Applying the Laplace method, we see that, asymptotically as $t\downarrow 0$,

$$\int_{\mathbb{R}}\left\{\frac{d(x)}{\sinh d(x)}\,\frac{\sinh d(0)}{d(0)}\right\}^{1/2}\exp\left[-\frac{(d(x)^2-d(0)^2)}{2t}\right]dx\sim I_2(t,n\ell(\gamma)),$$

so the assertion of Gutzwiller et al. follows. Roughly speaking, by combining the Van Vleck approximation and the Laplace method, they reduce the problem to an integration of Gaussian type and obtain the same representation of the second term in the right-hand side of (6.1). But, in this case, thanks to Ikeda and Matsumoto [**23**], we see that

$$\begin{aligned}&\int_{\mathbb{R}}p(t,x+\sqrt{-1}\,y,a^{2n}(x+\sqrt{-1}\,y))dx\\&\quad=y\int_W\left\{\int_{\mathbb{R}}\frac{1}{(2\pi V(t,w))^{1/2}}\exp\left[-\frac{(a^{2n}-1)^2x^2}{2V(t,w)}\right]dx\right\}\\&\qquad\times\delta_{\log a^{2n}}(w(t))P(dw),\end{aligned}\tag{6.2}$$

$$V(t,w)=\int_0^t\exp[2w(s)]ds,$$

and hence the integral is already of Gaussian type (of infinite dimension). Moreover, the value of the integral in the braces $\{\dots\}$ does not depend on $V(t,w)$, which is the only infinite dimensional quantity appearing in the integrand. Noting this fact, we can easily obtain the second term in the right-hand side of the first equation in (6.1). Furthermore, if we replace the w in $V(t,w)$ in (6.2) by the geodesic $c(s)$ connecting $\sqrt{-1}$ and $a^{2n}\sqrt{-1}$, and change the order of integration, it becomes clear why the same term as in the Selberg trace formula appears in the formula of Gutzwiller's semiclassical approximation. See [**23**] for details.

7. Supplement

(a) In [**34**], P. Lévy not only treated a prototype of (2.4) in Section 2.1, but explicitly calculated the inverse Fourier transform. Due to his result, the density $p(a)$ of the probabilty distribution of the stochastic area $A(t;w)$ of Brownian motion conditional on $w(t)=0$ is given by

$$p(a)=2\pi t\int_W\delta_{(0,a)}((w(t),A(t;w))P(dw)=\frac{1}{t(\cosh(2\pi a/t)+1)}.$$

By virtue of Lévy's formula (2.4), it follows that

$$\int_W \exp\left[\sqrt{-1}\,\lambda A(t;w)\right] P(dw) = \frac{1}{\cosh(\lambda t/2)}.$$

This is the case of $N = 0$ in (5.5), so we can prove the above formula in the same way as in Secton 3 without using Lévy's formula. Furthermore, it is known that the density $q(a)$ of the distribution of $A(t;w)$ is given by

$$q(a) = \int_W \delta_a(A(t;w))P(dw) = \frac{1}{t\cosh(\pi a/t)}$$

(see [**34**]).

(b) If we represent the left-hand side of (2.23) by using the eigenvalues of $B^{\#}$ and calculate the right-hand side by using the Jacobi fields, we obtain Euler's well-known formula

$$\prod_{n\in\mathbb{Z}, n\neq 0} \left\{1 - \frac{x}{2n\pi}\right\} \exp\left[\frac{x}{2n\pi}\right] = \frac{\sin(x/2)}{x/2}.$$

What we have done here, according to Kac [**31**] (p. 116), is to represent a quantity in two different ways: one is quantum mechanical and the other is classical mechanical, and the Van Vleck formula connects them. Performing analogous calculations on the action integral along a classical trajectory, we have a formula concerned with the double cosine function. For the details, see [**22**]. Not only in this case but in many concrete examples of quadratic Wiener functionals, we obtain formulas for $\det Q(z)$ using a canonical product of entire functons (see for example [**6**]).

(c) There are not many results discussing cases more general than the case stated in 2.7, but we refer to [**53**].

(d) It is known that not only does the stochastic area $A(t;w)$ play an important role as a typical example in quadatic forms, but it has a special position as a special function on W ([**24, 36**]). For example, the fact that $A(t;w)$ is discontinuous with respect to the uniform topology on W reflects on many results in stochastic analysis.

(e) There are several reasons why the Van Vleck formula for general Wiener integrals has been taken up. Concerning the background of the study of stochastic oscillatory integrals, P. Malliavin, who has contributed to this study since the 1970's, reported in [**37**]. Related to these, it is necessary to treat analogies to the stationary phase method and the method of steepest decent in Wiener space. See, for example, [**36, 37, 38, 52, 53**] and the literature cited in these references. Moreover, Malliavin and Taniguchi [**38**] studied a transformation formula in a complexified Wiener space, which turns out to be important in those studies, from a new point of view. For information about roles that the Van Vleck formula plays in the study of asymptotic properties of eigenvalues of the Schrödinger operator with an electromagneic field, see for example Matsumoto [**39**] and the literature therein.

(f) In the finite dimensional case, it is known that there exist phase functions in terms of which oscillatory integrals, not necessarily Gaussian, on special manifolds have explicit expressions. We only mention that there is some research concerning similar subjects on the path space.

References

[1] Ph. Biane - M. Yor; Variances sur une formule de Paul Lévy, Ann. Inst. H. Poincaré, Probab. Statist., **23** (1987), 359 – 377.

[2] A. N. Borodin - P. Salminen; Handbook of Brownian motion — Facts and formulae, Birkhäuser, 1996.
[3] R. H. Cameron - M. T. Martin; Transformations of Wiener integrals under translations, Ann. Math., **45** (1944), 386 – 396.
[4] R. H. Cameron - M. T. Martin; Evaluation of various Wiener integrals by use of certain Strum-Liouville differential equations, Bull. A.M.S., **51** (1945), 73 – 90.
[5] R. Cennni - E. Galleani d'Agllano - F. Napoli - P. Saracco-M. Sassetti; Feynman integrals in theoretical nuclear and statistical physics, Bibliopolis, 1989.
[6] T. Chan; Indefinite quadratic functionals of Gaussian processes and least-action paths, Ann. Inst. H. Poincaré Probab. Statist., **27** (1991), 239 – 271.
[7] T. Chan - D. S. Dean - K.M.Jansons - L. C. G. Rogers; On polymer conformations in elongational flows, Comm. Math. Phys., **160** (1994), 239 – 257.
[8] A. Comtet - B. Georgeot - S. Ouvry; Trace formula for Riemann surface with magnetic field, Phys. Review Lett., **71** (1973), 3786 – 3789.
[9] B. De Witt; Supermanifolds, Second Edition, Cambridge Univ. Press, 1992.
[10] C. De Witt-Morette; The semiclassical expansion, Ann. Phys., **97** (1976), 367 – 399.
[11] C. Donati-Martin - M. Yor; Fubini's theorem for double Wiener integrals and the variance of the Brownian path, Ann. Inst. H. Poincaré Probab. Statist., **27** (1991), 181 – 200.
[12] R. P. Feynman-A. R. Hibbs; Quantum mechanics and path integrals, McGraw-Hill, 1965.
[13] G. J. Foschini - L. A. Shepp; Closed form characteristic functions for certain random variables related to Brownian motion, in "Liber Amicorum for Moshe Zakai", Academic Press, 1991, 169 – 187.
[14] D.Fujiwara; Mathematical method of Feynman path integrals—Time division method (in Japanese), Springer Verlag Tokyo, 1999.
[15] C. Grosche - F. Steiner; Handbook of Feynman path integrals, Springer-Verlag, 1998.
[16] M. C. Gutzwiller; Chaos in classical and quantum mechanics, Springer-Verlag, 1990.
[17] M. C. Gutzwiller; Physics and Selberg's trace formula, Contemp. Math., **53** (1986), 215 – 251.
[18] L. Hörmander; The analysis of linear partial differential operators I, Springer-Verlag, 1983.
[19] N. Ikeda - S. Kusuoka - S. Manabe; Lévy's stochastic area formula for Gaussian processes, Comm. Pure Appl. Math., **47** (1994), 329–360.
[20] N. Ikeda - S. Kusuoka - S. Manabe; Lévy's stochastic area formula and related problems, in "Stochastic analysis", Proc. Symp. Pure Math. (Amer. Math. Soc.) **57** (1995), 281–305.
[21] N. Ikeda - S. Manabe; Asymptotic formulae for stochastic oscillatory integrals, in "Asymptotic problems in probability theory: Wiener functionals and asymptotics", Pitman Research Notes Math., **284** (1993), 136–155.
[22] N. Ikeda - S. Manabe; Van Vleck-Pauli formula for Wiener integrals and Jacobi fields, in "Itô's stochastic calculus and probability theory", Springer-Verlag, 1996, 141–156.
[23] N. Ikeda - H. Matsumoto; Brownian motion on the hyperbolic plane and Selberg trace formula, J. Funct. Anal., **163** (1999), 63 – 110.
[24] N. Ikeda - S. Watanabe; Stochastic differential equations and diffusion processes, 2nd ed., North-Holland/Kodan-sha, 1989.
[25] K. Itô - H. P. McKean; Diffusion processes and their sample paths, Springer, 1965.
[26] K. Itô - M. Nisio; On the convergence of sums of independent Banach space valued random variables, Osaka J. Math., **5** (1968), 35–48.
[27] C. J. Jacobi; Zur Theorie Variations-Rechnung und der Dieferential-Gleichungen, J. Reine Angew. Math., **17** (1837), 68 – 82.
[28] M. Kac; On some connections between probability theory and differential and integral equations, Proc. 2nd Berkeley Symp. on Math. Statist. and Probab., Univ. California Press, 1951, 189–215.
[29] M. Kac; Semi-classical quantum mechanics and Morse's theory, Trends in Appl. of Pure Math. to Mech., ed. by H. Zorski, Vol.2, 1979, 163–170.
[30] M. Kac; Integration in function spaces and some of its applications, Scoula Norumale Superiore, Pisa, 1980.
[31] M. Kac; Enigmas of chance; an autobiography, Harper & Row, 1985.
[32] D. C. Khandekar - S. V. Lawande - K. V. Bhagwat; Path-integral methods and their applications, World Scientific, 1993.

[33] S.Kotani; Measure and Probability, vol.2 (in Japanese), in Foundations of Modern Mathematics Series, Iwanami,1998.
[34] P. Lévy; Wiener's random function and other Laplacian random functions, Proc. 2nd Berkeley Symp. on Math. Statist. and Probab., Univ. California Press, 1951, 171–186.
[35] P. Malliavin; Stochastic calculus of variations and hypoelliptic operators, Proc. Intern. Symp. S. D. E., Kinokuniya, 1976, 195 – 263.
[36] P. Malliavin; Stochastic analysis, Springer-Verlag, 1997.
[37] P. Malliavin; Probability and geometry, Taniquchi Conf. Math. Nara '98, Adv. Stud. Pure Math. 31, Math. Soc. Japan, 2001, 179-209.
[38] P. Malliavin - S. Taniguchi; Analytic functions, Cauchy formula, and stationary phase on a real abstract Wiener space, J. Funct. Anal. **143** (1997), 470–528.
[39] H. Matsumoto; Semiclassical asymptotics of eigenvalues for Schrödinger operators with magnetic fields, J. Funct. Anal., **129** (1995), 719 – 756.
[40] H. Matsumoto; Quadratic Hamiltonians and associated orthogonal polynomials, J. Funct. Anal., **136** (1996), 214–225.
[41] H. Matsumoto - N. Ueki; Applications of the theory of the metaplectic representation of quadratic Hamiltonians on the two-dimensional Euclidean space, J. Math. Soc. Japan, **52** (2000), 269–292.
[42] H. P. McKean; Selberg's trace formula as applied to a compact surface, Comm. Pure Appl. Math., **25** (1972), 225 – 245.
[43] J. Milnor; Morse theory, Princeton Univ. Press, 1963.
[44] S. A. Molchanov, Diffusion processes and Riemannian geometry, Russian Math. Surveys, **30** (1975), no. 1, 1 – 63.
[45] C. Morette; On the definition and approximation of Feynman's path integral, Phys. Rev., **81** (1951), 848–852.
[46] R. E. C. Paley - N. Wiener; Fourier transforms in the complex domain, Colloq. Publ., **19**, Amer. Math. Soc., 1934.
[47] J. Rezende; Quantum systems with time-dependent harmonic part and the Morse index, J. Math. Phys., **25** (1984), 3264–3269.
[48] L. C. G. Rogers - Z. Shi; Quadratic functionals of Gaussian motion, optimal control, and the "Coldiz", Stochastic Rep., **41** (1992), 201 – 218.
[49] B.Sakita - K.Kikkawa; Quantum mechanics of many degrees of freedom by path integral methods (in Japanese), Iwanami, 1986.
[50] L. S. Schulman; Techniques and applications of path integration, John Wiley & Sons, 1981
[51] A. Selberg; Harmonic analysis and discontinuous groups in weakly symmetric Riemannian spaces with applications to Dirichlet series, J. Indian Math. Soc., **20** (1956), 47 – 87.
[52] H. Sugita - S. Taniguchi; Oscillatory integrals with quadratic phase function on a real abstract Wiener space, J. Funct. Anal., **155** (1998), 229 – 262.
[53] S. Taniguchi; Stochastic oscillatory integrals with quadratic phase function and Jacobi equations, Probab. Theory Relat. Fields, **114** (1999), 291 – 308.
[54] J. H. Van Vleck; The correspondence principle in the statistical interpretation of quantum mechanics, Proc. Nat. Acad. Sci. U.S.A., **14** (1928), 178–188.
[55] S. Watanabe; Lectures on stochastic differential equations and Malliavin calculus, Tata Inst. Fund. Research, Springer-Verlag, 1984.
[56] M. Yor; On stochastic areas and averages of planar Brownian motion, J. Phys. A Math. Gen., **22** (1989), 3009 – 3057.

SUZUKAKEDAI 2-40-3, SANDA, HYOGO 669-1322, JAPAN

Translated by SHOJIRO MANABE

Amer. Math. Soc. Transl.
(2) Vol. **211**, 2003

Spectral Geometry for Schrödinger Operators in a Magnetic Field Focusing on Geometrical and Dynamical Structures on Manifolds

Ruishi Kuwabara

Introduction

A second order differential operator $\hat{H}$ on an n-dimensional smooth connected manifold M is locally expressed as

$$\hat{H} = \sum_{j,k=1}^{n} p^{jk}(x)\frac{\partial^2}{\partial x^j \partial x^k} + \sum_{j=1}^{n} q^j(x)\frac{\partial}{\partial x^j} + r(x) \quad (p^{jk}(x) = p^{kj}(x)) \tag{0.1}$$

using local coordinates $(x^1, \ldots, x^n)$. In order for the operator $\hat{H}$ to be globally defined on M, the coefficients p^{jk}, q^j and r must be changed to

$$p'^{jk} = \sum_{l,m} p^{lm}\frac{\partial x'^j}{\partial x^l}\frac{\partial x'^k}{\partial x^m}, \quad q'^j = \sum_l q^l \frac{\partial x'^j}{\partial x^l} + \sum_{l,m} p^{lm}\frac{\partial^2 x'^j}{\partial x^l \partial x^m}, \quad r' = r,$$

under the change of coordinates $(x^j) \to (x'^j)$. Here the first equation shows that the p^{jk} constitute a symmetric contravariant 2-tensor field on M. Hence, if we assume $\hat{H}$ to be a positive elliptic operator, we have a Riemannian metric $g = (g_{jk}) := -(p^{jk})^{-1}$ on M. Let ∇ denote the Levi-Civita connection (the covariant derivative) induced from the Riemannian metric g. Then, (0.1) is rewritten as

$$\hat{H} = -\sum_{j,k} g^{jk}(x)\nabla_j\nabla_k + \sum_j Q^j(x)\nabla_j + r(x) \quad \Big(Q^j(x) := q^j(x) - \sum_{l.m} g^{lm}(x)\Gamma^j_{lm}(x)\Big)$$

(Γ^j_{lm} is the Christoffel symbol), where $Q^j(x)$ turns out to be a (complex-valued) vector field on M. This gives a global expression of $\hat{H}$.

Let us take the natural inner product of $C_0^\infty(M)$ (the space of C^∞ functions with compact support) induced from the Riemannian metric g. The operator $\hat{H}$ is

This article originally appeared in Japanese in Sūgaku **54** (1) (2002), 37–57.
2000 *Mathematics Subject Classification.* Primary 58Jxx; Secondary 35Q40.

symmetric with respect to this inner product if and only if it is in the form

(0.2)
$$\hat{H} = -\sum_{j,k} g^{jk}(x)\{\nabla_j - iA_j(x)\}\{\nabla_k - iA_k(x)\} + V(x) \quad \Big(A_j(x) := \sum_k g_{jk}(x)A^k(x)\Big)$$

($i := \sqrt{-1}$) with $A^j(x)$ being a real vector field and $V(x)$ a real-valued function on M. As a consequence, we have a one-to-one correspondence between second order, elliptic, symmetric differential operators $\hat{H}$ and the geometric structures on M: (g, A, V), the triple of Riemannian structures, real vector fields and real-valued functions.

From the dynamical viewpoint we regard $\hat{H}$ as a Schrödinger operator, and by the correspondence rule $-i\nabla_j \mapsto \xi_j$ we have the Hamiltonian function

$$H(x,\xi) = \sum_{j,k} g^{jk}(x)\{\xi_j - A_j(x)\}\{\xi_k - A_k(x)\} + V(x) \tag{0.3}$$

for the classical dynamical system. Here, $(x, \xi) = (x^1, \ldots, x^n, \xi_1, \ldots, \xi_n)$ can be regarded as local coordinates of the cotangent bundle T^*M of M (i.e., $T^*M \ni p = \sum_j \xi_j dx^j$), and H is a real-valued function on T^*M. On T^*M there is naturally defined the symplectic structure $\Omega_M = \sum_j d\xi_j \wedge dx^j$, and we get the Hamiltonian dynamical system $\mathcal{H}_{cl}$, in which the orbits are governed by the canonical equation

(0.4)
$$\begin{cases} \dot{x}^j &= \dfrac{\partial H}{\partial \xi_j} = 2\sum_k g^{jk}\xi_k - 2A^j, \\ \dot{\xi}_j &= -\dfrac{\partial H}{\partial x^j} = -\sum_{l,m} \dfrac{\partial g^{lm}}{\partial x^j}\xi_l\xi_m + 2\sum_k \dfrac{\partial A^k}{\partial x^j}\xi_k - \dfrac{\partial}{\partial x^j}\Big(\sum_k A_k A^k\Big) - \dfrac{\partial V}{\partial x^j}. \end{cases}$$

This is equivalent to the second order equation on M

(0.5)
$$\ddot{x}^j + \sum_{l,m} \Gamma^j_{lm}\dot{x}^l\dot{x}^m - 2\sum_{l,m} g^{jl}\Theta_{lm}\dot{x}^m + 2\sum_k g^{jk}\frac{\partial V}{\partial x^k} = 0 \quad \Big(\Theta_{lm} := \frac{\partial A_m}{\partial x^l} - \frac{\partial A_l}{\partial x^m}\Big),$$

which describes the motion of a charged particle (with unit charge) on (M, g) under the electric field V and the magnetic field Θ_{lm}. It is just the equation of geodesics on (M, g) when $A^j = V = 0$. We remark here that (being different from the quantum system $\hat{H}$) the classical dynamical system $\mathcal{H}_{cl}$ does not directly depend on $\alpha := \sum A_j dx^j$, but depends only on $\Theta := d\alpha = \frac{1}{2}\sum \Theta_{lm} dx^l \wedge dx^m$.

The purpose of so-called "spectral geometry" is to study the relationships among (i) the analytic structures, especially the spectrum, of the operator $\hat{H}$, (ii) the geometric invariants of the triple (g, A, V), and (iii) the dynamical properties of the classical system $\mathcal{H}_{cl} = (T^*M, \Omega_M, H)$. As is well known, studies (in the case of $A = V = 0$) on the relationship between the spectrum of $\hat{H} = \Delta_g$ (the Laplace-Beltrami operator) and the geometric invariants of (M, g) have been widely developed from various viewpoints, starting from work by McKean and Singer [**27**], Berger-Gauduchon-Mazet [**2**], and so on since the late 1960s, and a large number of results have been accumulated. From the dynamical viewpoint we can point out the trace formula, which clarifies the relation between the spectrum and the geodesics (the geodesic flow). Furthermore, there are a huge number of physics articles dealing with the connection between the quantum and classical systems (in Euclidean spaces).

The aim of this article is to present some topics and results on the relationships among the spectrum of $\hat{H}$, geometry of (M, g, A), and mechanics in $\mathcal{H}_{cl}$ in the case where M is a compact manifold without boundary and the scalar potential $V = 0$. In the framework of symplectic structures and their quantized objects we will consider how the magnetic field Θ or its vector potential A affects the quantum spectrum from the geometric and dynamical viewpoints. The techniques used here are based on the theory of Hamiltonian mechanics, the geometry of connections, and the theory of pseudodifferential and Fourier integral operators (refer to [**1**], [**16**], [**17**], [**15**], [**33**], for example). Let us mention that the main results in this survey were obtained around 1990, and we cannot cover many arguments that have recently been deeply and widely developed on this subject.

1. Dynamical systems in a magnetic field

1.1. Classical magnetic flows and Schrödinger operators on a line bundle. In the previous section we called the (exact) two-form $\Theta = d\alpha$ a magnetic field with the magnetic potential α (which is a one-form on M). In a generalized setting we call a (probably non-exact) closed two-form Θ a *magnetic field*, and we have the equation (0.5) (with $V \equiv 0$) describing the motion of a charged particle in the magnetic field Θ on (M, g). In this case there does not exist an associated vector potential A^j globally defined on M, and so the Hamiltonian (0.3) is not defined as a function on T^*M. However, we can introduce a Hamiltonian system on T^*M as follows: By means of the magnetic field Θ let us define the twisted symplectic form $\Omega := \Omega_M + \pi_M^*\Theta$ $(\pi_M : T^*M \to M)$ on T^*M. Taking $H_0(x,\xi) = \sum g^{jk}(x)\xi_j\xi_k$ as the Hamiltonian function, we have a Hamiltonian system $\mathcal{H}_{\mathrm{mag}} = (T^*M, \Omega, H_0)$, in which the canonical equation is given by

$$
\left\{
\begin{array}{lcl}
\dot{x}^j & = & 2\sum\limits_k g^{jk}\xi_k, \\
\dot{\xi}_j & = & -\sum\limits_{l,m} \dfrac{\partial g^{lm}}{\partial x^j}\xi_l\xi_m - 2\sum\limits_{l,m} g^{lm}\Theta_{lj}\xi_m.
\end{array}
\right.
\tag{1.1}
$$

The flow governed by this equation (which is found to be equivalent to (0.5)) is called a *magnetic flow* on T^*M.

Next, we introduce the quantum system associated to the classical system $\mathcal{H}_{\mathrm{mag}}$. In this procedure we need the following condition ("quantization condition") for the magnetic field Θ:

$$
\text{(H)} \qquad [\Theta/2\pi] \in H^2_D(M, \boldsymbol{Z}),
$$

where $[\cdot]$ denotes the cohomology class $(\in H^2_D(M, \boldsymbol{R}))$ and $H^2_D(M, \boldsymbol{Z})$ is the image of the map $H^2(M, \boldsymbol{Z}) \to H^2(M, \boldsymbol{R}) \cong H^2_D(M, \boldsymbol{R})$ induced from the inclusion $\boldsymbol{Z} \to \boldsymbol{R}$. Under the assumption (H), the Chern-Weil-Kostant theory [**18**] asserts that there uniquely exists a complex line bundle $\pi_E : E \to M$ over M whose Chern class is $[\Theta/2\pi]$. Furthermore, there exist on E a connection $\widetilde{\nabla}$ whose curvature is $-i\Theta$ and an Hermitian inner product compatible with $\widetilde{\nabla}$. Let $C^\infty(E)$ be the space of smooth sections of E, and let $\Lambda^p(E)$ be the space of E-valued smooth p-forms on M. The connection $\widetilde{\nabla}$ is a linear map $\widetilde{\nabla} : C^\infty(E) \to \Lambda^1(E)$ that is locally expressed as follows. Take a local section $e : U(\subset M) \to E$ such that $\|e(x)\| = 1\,(x \in U)$. Then, for $s \in C^\infty(E|_U)$ written as $s(x) = f(x)e(x)$ $(f(x) \in C^\infty(U))$ we have

$$\widetilde{\nabla} s = (df - if\alpha)e,$$

where α is a one-form on U satisfying $d\alpha = \Theta$. The purely imaginary one-form $-i\alpha$ is called the *connection form* of $\widetilde{\nabla}$ with respect to the local section e. From the Riemannian metric on M and the Hermitian metric on E we can naturally define the inner products for elements of $C^\infty(E)$ and $\Lambda^p(E)$ with compact support. With respect to this inner product let $\widetilde{\nabla}^* : \Lambda^1(E) \to C^\infty(E)$ be the adjoint operator of $\widetilde{\nabla}$. We regard the operator $L := -\widetilde{\nabla}^*\widetilde{\nabla}$ (called the *Bochner-Laplacian*) acting on $C^\infty(E)$ as the *Schrödinger operator* corresponding to the classical system $\mathcal{H}_{\mathrm{mag}}$. For a local section $s = fe \in C^\infty(E|_U)$ we have

$$Ls = \Big[-\sum_{j,k} g^{jk}\{\nabla_j - iA_j\}\{\nabla_k - iA_k\}f\Big]e \quad \Big(\alpha = \sum_j A_j dx^j\Big). \tag{1.2}$$

Each state for the quantum system is an element of the space $L^2(E)$ of L^2-sections of E, and the probability for a particle to be found in the region D is given by $\int_D \|s(x)\|^2 d\mu_g$ $(d\mu_g := \sqrt{g}dx^1 \cdots dx^n)$.

REMARK. If $H^1(M, \boldsymbol{R}) \neq \{0\}$, the connection $\widetilde{\nabla}$ for the magnetic field Θ is not unique (see §2.1). In this case, for the classical system uniquely determined by Θ there exist different Schrödinger operators, whose spectra may be different. This phenomenon is known as the Aharanov-Bohm effect.

1.2. Dynamical system on a principal $U(1)$-bundle and its reduction. We here give another formulation of the classical and the quantum systems using a principal $U(1)$-bundle over M. This formulation is useful to analyze the spectrum (or eigenfunctions) of the Schrödinger operators.

Let $\pi : P \to M$ be the principal $U(1)$-bundle associated to the Hermitian line bundle E; more precisely, $E = P \times_\rho \boldsymbol{C} := (P \times \boldsymbol{C})/\sim$, where ρ is the unitary representation of $U(1) = \{e^{i\theta}; 0 \le \theta < 2\pi\}$ on $\boldsymbol{C}$ defined by $\rho(e^{i\theta})(z) = e^{-i\theta}z$, and the equivalence relation in $P \times \boldsymbol{C}$ is defined by $(p, z) \sim (p \cdot e^{i\theta}, \rho(e^{-i\theta})z)$. The connection $\widetilde{\nabla}$ on E naturally induces the connection on P (denoted by the same symbol $\widetilde{\nabla}$). The connection $\widetilde{\nabla}$ on P gives the decomposition (invariant under the $U(1)$-action) of the tangent space $T_pP = V_p \oplus H_p$ at each point $p \in P$, where V_p is the tangent space of the fiber (called the vertical space) and H_p is called the horizontal space characterizing $\widetilde{\nabla}$. Take an invariant metric on $U(1)$ (such that $\|\partial/\partial\theta\| = c > 0$), and we can uniquely introduce a Riemannian metric $\tilde{g}$ (called the *Kaluza-Klein metric*) on P from the metric g on M and the connection $\widetilde{\nabla}$ on P that satisfies (i) the metric on the fiber is that on $U(1)$, (ii) $\pi_* : H_p \to T_{\pi(p)}M$ is an isometry, and (iii) V_p and H_p are orthogonal.

We have the system of geodesic flow $\widetilde{\mathcal{H}}_0 = (T^*P, \Omega_P, \widetilde{H}_0)$ associated to the metric $\tilde{g}$ on P. The action of the structure group $U(1)$ on P is naturally lifted to the symplectic action on (T^*P, Ω_P), that is, an action preserving Ω_P. Furthermore, this action also preserves the Hamiltonian function $\widetilde{H}_0$. In this situation the Marsden-Weinstein reduction procedure [**26**] goes well, as follows. First we have the moment map $J : T^*P \to \mathfrak{u}(1)^*$ associated to the $U(1)$-action, where $\mathfrak{u}(1)^* = \{\tau d\theta;\ \tau \in \boldsymbol{R}\}$ is the dual space of the Lie algebra $\mathfrak{u}(1)$ of $U(1)$. Next, for each $\mu \in \boldsymbol{R} \cong \mathfrak{u}(1)^*$, the subset $J^{-1}(\mu)$ $(\subset T^*P)$ is invariant under the $U(1)$-action, and is an invariant submanifold of the dynamical system $\widetilde{\mathcal{H}}_0$. Since the flow on $J^{-1}(\mu)$ commutes with the action of $U(1)$, we obtain an induced dynamical system on the quotient space $P_\mu := J^{-1}(\mu)/U(1)$. Finally, this dynamical system is formulated as a Hamiltonian system $\mathcal{P}_\mu = (P_\mu, \Omega_\mu, H_\mu)$, where the symplectic form Ω_μ is naturally induced on P_μ from Ω_P on T^*P, and the Hamiltonian H_μ is induced from the $U(1)$-invariant function $\widetilde{H}_0$. Thus we have a family of induced dynamical systems $\mathcal{P}_\mu$ ($\mu \in \boldsymbol{R}$). The relationship between the systems $\mathcal{P}_\mu$ and the magnetic flow on T^*M is understood as follows.

PROPOSITION 1.1. *For each $\mu \in \boldsymbol{R} \cong \mathfrak{u}(1)^*$ there exists a diffeomorphism $\Psi_\mu : P_\mu \to T^*M$ such that*

$$\Omega_\mu = \Psi^*\{\Omega_M + \pi_M^*(\mu\Theta)\}, \quad H_\mu = \Psi^* H_0 + |\mu|^2/c^2.$$

In particular, $\mathcal{H}_{\mathrm{mag}} = (T^*M, \Omega, H_0)$ is isomorphic with $\mathcal{P}_1 = (P_1, \Omega_1, H_1)$ as Hamiltonian systems, and $\mathcal{P}_0$ is isomorphic with the system of geodesic flow on T^*M.

Let us express the above discussion by using local coordinates. Let $\bar{x} = (x, \theta)$ ($x \in U, \theta \in [0, 2\pi)$) be local coordinates of a local coordinate neighborhood $\pi^{-1}(U) \cong U \times U(1)$ of P, and let $(\bar{x}, \bar{\eta}) = (x, \theta, \eta, \tau)$ be the local canonical coordinates of T^*P. The Hamiltonian function $\widetilde{H}_0$ is expressed as

$$\widetilde{H}_0(\bar{x}, \bar{\eta}) = \sum g^{jk}(x)\eta_j\eta_k - 2\sum g^{jk}(x)A_k(x)\eta_j\tau + \big(\|A(x)\|^2 + \frac{1}{c^2}\big)\tau^2.$$

Note here that the horizontal space H_p of $\widetilde{\nabla}$ is generated by $\{\partial/\partial x^j - A_j(x)\partial/\partial\theta\}_{j=1}^n$ for $A_j(x)$ with $d(\sum A_j(x)dx^j) = \Theta$. The canonical equation (the geodesic flow) on T^*P is

$$(1.3) \qquad \begin{cases} \dot{x}^i = 2(\sum g^{ij}\eta_j - A^i\tau), \\ \dot{\eta}^i = -\sum \dfrac{\partial g^{jk}}{\partial x^i}\eta_j\eta_k + 2\sum \dfrac{\partial A^j}{\partial x^i}\eta_j\tau - \dfrac{\partial}{\partial x^i}(\|A\|^2)\tau^2, \\ \dot{\theta} = -2\sum A^j\eta_j + 2(\|A\|^2 + \dfrac{1}{c^2})\tau, \\ \dot{\tau} = 0 \quad (\text{hence, } \tau = \text{Const}). \end{cases}$$

The submanifold $J^{-1}(\mu)$ of $T^*P = \{(x, \theta, \eta, \tau)\}$ is determined by $\tau = \mu$, and we can take (x, η) as local coordinates of P_μ. Then, using the canonical coordinates (x, ξ) of T^*M, we can express the diffeomorphism Ψ_μ as $(x^j, \eta_j) \mapsto (x^j, \xi_j) = (x^j, \eta_j - \mu A_j(x))$.

In the system $\mathcal{P}_0$ each solution $x(t)$ of (1.3) is a geodesic on (M, g), and $(x(t), \theta(t))$ is a horizontal curve with respect to the connection $\widetilde{\nabla}$ on P. Let $\gamma(t)$ be a periodic curve on M with period T and let $\tilde{\gamma}(t)$ ($\in P$) be a horizontal lift of γ. Then, $\tilde{\gamma}(T) = \tilde{\gamma}(0) \cdot \lambda$ holds for some $\lambda \in U(1)$, and we call λ the *holonomy* of $\widetilde{\nabla}$ along γ.

Next, let us consider a "reduction" of the quantum system. Let Δ_P be the Laplace-Beltrami operator on $(P, \tilde{g})$. For each integer m, consider the unitary representation $\rho_m(e^{i\theta})(z) := e^{-im\theta}z$ of $U(1)$ on $\boldsymbol{C}$. A function f on P satisfying $f(p \cdot e^{i\theta}) = e^{-im\theta}f(p)$ ($p \in P$) is said to be *ρ_m-equivariant*, and the spaces of such functions are denoted by $C_m^\infty(P)$, $L_m^2(P), \ldots$. Noting the Fourier series expansion,

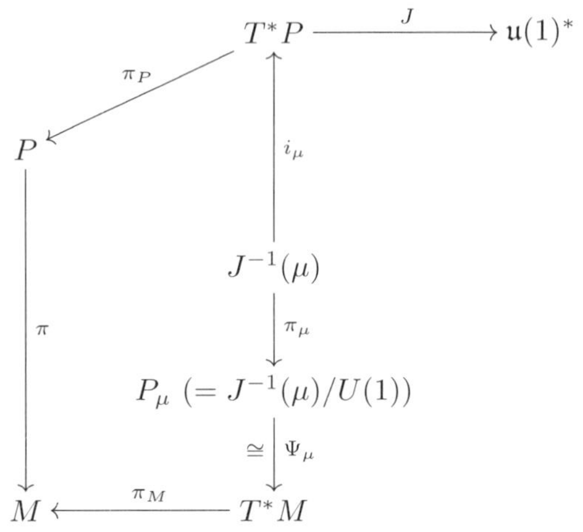

FIGURE 1. Reduction of the dynamical system

we have the decomposition

$$L^2(P) = \bigoplus_{m\in \boldsymbol{Z}} L^2_m(P)\,. \tag{1.4}$$

Since the $U(1)$-action gives isometries on $(P, \tilde{g})$, Δ_P leaves each $L^2_m(P)$ invariant. Hence, we define $\widetilde{L}_m := \Delta_P|_{L^2_m(P)}$.

Let $E_m = P \times_{\rho_m} \boldsymbol{C}$ be the associated line bundle of P by the representation ρ_m. Let $\widetilde{\nabla}^{(m)}$ be the connection on E_m induced from the connection $\widetilde{\nabla}$ on P. Then the curvature of $\widetilde{\nabla}^{(m)}$ is equal to $-im\Theta$. Let L_m be the Bochner-Laplacian on E_m, which is locally expressed as

$$L_m s = \Big[-\sum_{j,k} g^{jk}\{\nabla_j - imA_j\}\{\nabla_k - imA_k\}f\Big]e_m \quad (s = fe_m \in C^\infty(E_m|_U)), \tag{1.5}$$

where the local section e_m of E_m is defined by $e_m(x) = [(p,1)]_m \in P\times_{\rho_m}\boldsymbol{C}$ ($p \in P, x = \pi(p)$) with $e(x) = [(p,1)] \in P\times_\rho \boldsymbol{C}$ being the local section in the formula (1.2). We have the unitary isomorphism

$$\psi_m : L^2_m(P) \to L^2(E_m)\,;\ \ f \mapsto s, \tag{1.6}$$

with $s(x) = [(p, f(p))]_m$ ($\pi(p) = x$). Then,

PROPOSITION 1.2. *We have*

$$\widetilde{L}_m = \psi_m^{-1} L_m \psi_m + |m|^2/c^2.$$

In particular, the Schrödinger operator L $(= L_1)$ *is unitarily equivalent to* $\widetilde{L}_1$ *(without the difference of the constant term), and the Laplace-Beltrami operator* Δ_M $(= L_0)$ *on* (M, g) *is unitarily equivalent to* $\widetilde{L}_0$.

Thus, we see that the Hamiltonian system $\mathcal{H}^{(\mu)}_{\mathrm{mag}} := (T^*M, \Omega_M + \pi^*_M(\mu\Theta), H_0) \cong \mathcal{P}_\mu$ describes the motion of a particle with charge $\mu \in \boldsymbol{R}$ in the magnetic field Θ, and that if the quantization condition $\mu = m \in \boldsymbol{Z}$ for the charge is satisfied, there exists the corresponding Schrödinger operator L_m.

On the other hand, let us consider the following operator obtained from (1.5) divided by m^2:

$$\hat{H}_m = -\sum_{j,k} g^{jk}\Big\{\frac{1}{m}\nabla_j - iA_j\Big\}\Big\{\frac{1}{m}\nabla_k - iA_k\Big\}. \tag{1.7}$$

Here we regard $\hbar := 1/m$ as the Planck constant. Then, the consideration of the spectrum of L_m for $m \to \infty$ is just the so-called *semi-classical analysis* of the Schrödinger operator $\hat{H}_m$ (for $1/m = \hbar \ll 1$) associated to the classical system $\mathcal{H}_{\mathrm{mag}}$ (refer to the discussions in §4).

2. Quantum spectra in magnetic fields: Basic properties and examples

From now on we assume that M is compact (that was not assumed in the previous section). Then, it is known that for each $m \in \boldsymbol{Z}$ the operator L_m whose domain is $C^\infty(E_m)$ ($\in L^2(E_m)$) is essentially self-adjoint, and that the resolvent $(L_m - zI)^{-1}$ is a compact operator. Hence the spectrum of (the closure of) L_m (which is non-negative and is unbounded) consists of real eigenvalues with finite multiplicity, which are denoted by

$$(0 \le)\ \lambda_1^{(m)} \le \lambda_2^{(m)} \le \cdots \le \lambda_j^{(m)} \le \cdots \uparrow \infty\,.$$

REMARK. In the case where M is not compact, it is known that L_m is essentially self-adjoint if M is complete. However, the spectrum does not necessarily consist of eigenvalues.

We will investigate in this article the spectrum $\{\lambda_j^{(m)}\}$ (which is denoted by $\mathrm{Sp}(M, g; E_m, \widetilde{\nabla}^{(m)})$) in the presence of the magnetic field, paying attention to the relationship with the geometric structure $(M, g; E_m, \widetilde{\nabla}^{(m)})$ and the dynamical properties of the system $\mathcal{H}_{\mathrm{mag}}^{(m)} = (T^*M, \Omega_M + \pi_M^*(m\Theta), H_0)$.

We point out that there are two directions in studying $\{\lambda_j^{(m)}\}$. The first is the following: We fix m ($= 1$, for example), and investigate the spectrum $\{\lambda_j^{(1)}\}$ of the Bochner-Laplacian L in connection with the geometry of $(M, g; E, \widetilde{\nabla})$. This is a generalization of the studies on the relationship between the spectrum of the Laplace-Beltrami operator Δ_g and the geometric invariants (the curvatures, the geodesics, for example) of (M, g). This research is to clarify how the first-order term of L (the vector potential of Θ) affects the spectrum of Δ_g. For this purpose we have only to analyze L on the line bundle E without introducing P. Another direction of research is from the dynamical viewpoint, more precisely, to investigate the spectrum in connection with the dynamical system $\mathcal{H}_{\mathrm{mag}} = (T^*M, \Omega, H_0)$. In this approach it will turn out that the dynamical properties of $\mathcal{H}_{\mathrm{mag}}$ are closely related to the spectrum $\{\lambda_j^{(m)}\}$ for integers $m \in \boldsymbol{Z}$ (or the asymptotic distribution as $m \to \infty$), and that the formulation based on the reductions of the system on P plays an essential role.

2.1. Basic properties. We review basic properties of the spectrum $\{\lambda_j^{(m)}\}$.

The gauge transformation. Let E be a Hermitian line bundle over M. A diffeomorphism $\varphi : E \to E$ is called a *gauge transformation* of E if it induces a linear isometry of each fiber E_x $(x \in M)$. For a connection $\widetilde{\nabla}$ we can define another connection $\varphi^*\widetilde{\nabla}$ by $(\varphi^*\widetilde{\nabla})s := \varphi^{-1}(\widetilde{\nabla}(\varphi \circ s))$ $(s \in C^\infty(E))$. Let $-i\alpha$ and $-i\alpha'$ be the connection forms (relative to the local section e) of $\widetilde{\nabla}$ and $\varphi^*\widetilde{\nabla}$, respectively. Then, it turns out that

$$-i(\alpha' - \alpha) = \phi^{-1} d\phi \quad (\varphi(e(x)) = \phi(x)e(x)), \tag{2.1}$$

and that the curvatures of these connections are the same. Note that (2.1) gives a global relation on M; that is, $\phi(x)$ is a function on M not depending on the choice of e. We say two connections $\widetilde{\nabla}$ and $\varphi^*\widetilde{\nabla}$ are *gauge equivalent*.

Suppose we are given a curvature form $-i\Theta$. Then, the set of the connections (the connection forms) of curvature $-i\Theta$ is given by $\{-i(\alpha_0 + \beta);\ \beta \in C\Lambda^1(M)\}$, where α_0 is a connection form such that $d\alpha_0 = \Theta$, and $C\Lambda^1(M)$ stands for the space of real-valued closed one-forms on M. Noticing (2.1), we see that

$$-i\alpha' \text{ and } -i\alpha \text{ are gauge equivalent} \iff [(\alpha' - \alpha)/2\pi] \in H^1_D(M, \boldsymbol{Z}).$$

As a consequence, it turns out that the quotient space of connections (with the curvature $-i\Theta$) divided by the gauge equivalence relation is identified with the torus $H^1(M, \boldsymbol{R})/H^1_D(M, \boldsymbol{Z})$.

We can easily see the following fact (concerning the first eigenvalue).

FACT 2.1. *For* $\mathrm{Sp}(M, g; E_m, \widetilde{\nabla}^{(m)}) = \{\lambda_j^{(m)};\ j \in \boldsymbol{N}\}$ *the following conditions are equivalent to each other:*

(1) For some (then, all) $m \neq 0$, *we have* $\lambda_1^{(m)} = 0$.

(2) For some (then, all) $m \neq 0$ *and any* $j \in \boldsymbol{N}$, *we have* $\lambda_j^{(m)} = \lambda_j^{(0)}$.

(3) For some (then, all) $m \neq 0$, E_m *is the trivial bundle with the connection* $\widetilde{\nabla}^{(m)}$ *gauge equivalent to the connection* $\widetilde{\nabla}_0$ *whose connection form is identically zero.*

When the above conditions are satisfied, the curvature $-im\Theta$ *of* $\widetilde{\nabla}^{(m)}$ *is identically zero. If* M *satisfies* $H^1(M, \boldsymbol{R}) = \{0\}$, *then the condition* $\Theta \equiv 0$ *is equivalent to the conditions (1), (2), (3).*

Isospectral problems. Obviously, $\mathrm{Sp}(M, g; E, \widetilde{\nabla}) = \mathrm{Sp}(M, g; E, \varphi^*\widetilde{\nabla})$. So we are interested in the inverse problem (the *isospectral problem*):

Does $\mathrm{Sp}(M, g; E_1, \widetilde{\nabla}_1) = \mathrm{Sp}(M, g; E_2, \widetilde{\nabla}_2)$ imply that $E_1 \cong E_2$ and the connections $\widetilde{\nabla}_1$ and $\widetilde{\nabla}_2$ are gauge equivalent?

Concerning this problem we have the following.

FACT 2.2 ([**21**], [**22**]). *(1) On a line bundle over a certain Riemannian manifold (a hyperbolic surface, for example) there exist two isospectral connections with different curvatures.*

(2) On a line bundle over a certain nilmanifold there exists a one-parameter family of isospectral connections that are not gauge equivalent.

Fact (1) in the above is found by generalizing Sunada's theory [**30**], and (2) is found by an idea due to Gordon and Wilson [**9**].

Minakshisundaram-Pleijel expansion. The trace of the "heat operator" $\exp(-tL_m)$ $(t > 0)$ for L_m has the asymptotic expansion as $t \downarrow 0$ (the so-called Minakshisundaram-Pleijel expansion)

$$\sum_{j=1}^{\infty} \exp(-\lambda_j^{(m)} t) \sim (4\pi t)^{-n/2}(a_0^{(m)} + a_1^{(m)} t + a_2^{(m)} t^2 + \cdots).$$

Here the coefficients $a_0^{(m)}$, $a_1^{(m)}, \ldots$ are given by

$$a_0^{(m)} = \mathrm{vol}(M, g), \qquad a_1^{(m)} = \frac{1}{6}\int_M \tau d\mu_g,$$

$$a_2^{(m)} = \frac{1}{360}\int_M (2\|R\|^2 - 2\|\rho\|^2 + 5\tau^2) d\mu_g - \frac{m^2}{12}\int_M \|\Theta\|^2 d\mu_g,$$

where R, ρ, and τ are the curvature tensor, the Ricci tensor, and the scalar curvature of (M, g), respectively. The general term $a_k^{(m)}$ is known to be given as the integral of a scalar on M that is obtained by the contraction of a polynomial in R, ρ, τ, Θ and their derivatives. However, it is hard to obtain the explicit form of $a_k^{(m)}$. Noticing the formula for $a_2^{(m)}$, we have the following.

FACT 2.3. *If* $\mathrm{Sp}(M, g; E_1, \widetilde{\nabla}_1) = \mathrm{Sp}(M, g; E_2, \widetilde{\nabla}_2)$ *and* $\widetilde{\nabla}_1$ *is a flat connection, then* $E_2 \cong E_1$ *(the trivial bundle), and* $\widetilde{\nabla}_2$ *is also a flat connection.*

The spectrum of Δ_P. Let $\mathrm{Sp}(P, \tilde{g})$ denote the spectrum of the Laplace-Beltrami operator Δ_P on P endowed with the Kaluza-Klein metric. Then, from Proposition 1.2 and (1.4) we get

FACT 2.4.

$$\mathrm{Sp}(P, \tilde{g}) = \bigcup_{m \in \boldsymbol{Z}} \{\lambda_j^{(m)} + \frac{|m|^2}{c^2};\ j \in \boldsymbol{N}\}.$$

This fact plays a useful role in computing the spectrum of L_m (see examples below).

2.2. Examples of magnetic fields and their spectra.

2.2.1. *The zero magnetic field on a flat torus.* Let us consider the magnetic field $\Theta \equiv 0$ on a flat torus $\boldsymbol{T}^n := \Gamma\backslash\boldsymbol{R}^n$ (Γ is a lattice in $\boldsymbol{R}^n$) of dimension n. The line bundle associated to Θ is trivial, and it is easy to see that the connection forms with zero curvature are given in the form $-i\alpha = -i\sum_j a_j dx^j$, where the a_j are real constants. Let Γ^* be the dual lattice of Γ, i.e., $\Gamma^* = \{\xi \in \boldsymbol{R}^n;\ \xi \cdot \gamma \in \boldsymbol{Z}$ for all $\gamma \in \Gamma\}$. Then, the spectrum of the Bochner-Laplacian L_α associated to the connection form $-i\alpha$ is given by

$$\{\|2\pi\xi - a\|^2;\ \xi \in \Gamma^*\} \qquad (a := (a_1, \ldots, a_n)),$$

the corresponding eigenfunctions being $\{s_\xi(x) := \exp(2\pi i x \cdot \xi)\}$. Thus there exist different connections (vector potentials) associated to the zero magnetic field, and the corresponding quantum systems have different energy levels (the Aharanov-Bohm effect).

2.2.2. *A uniform magnetic field on a two-dimensional flat torus.* Consider the Heisenberg group

$$H_1 := \left\{(x, y, z) := \begin{pmatrix} 1 & x & z \\ 0 & 1 & y \\ 0 & 0 & 1 \end{pmatrix}; \; x, y, z \in \boldsymbol{R}\right\}$$

and its discrete subgroup

$$\Gamma := \{(x, y, z) \in H_1; \; x, y, z \in \boldsymbol{Z}\}.$$

We have the quotient manifold $P = \Gamma \backslash H_1$ (called a nilmanifold). The group $S^1 = \{(0, 0, z) \in H_1; \; 0 \leq z < 1\}$ acts on P on the right, and the quotient space P/S^1 by this action is diffeomorphic with the two-dimensional torus $\boldsymbol{T}^2$. Thus we have the principal $U(1)$-bundle $\pi : P \to \boldsymbol{T}^2$; $[(x, y, z)] \mapsto [(x, y)]$.

There exist three left-invariant vector fields

$$e_1 := \partial/\partial x, \quad e_2 := \partial/\partial y + x\partial/\partial z, \quad e_3 := \partial/\partial z$$

on H_1. We introduce the left-invariant metric on H_1 relative to which these vector fields form an orthonormal system, and let $\tilde{g}$ be the induced metric on P. Take the metric on $\boldsymbol{T}^2$ such that the vector fields $\pi_*(e_1) = \partial/\partial x$ and $\pi_*(e_2) = \partial/\partial y$ form an orthonormal system. Then, it turns out that this is a flat metric on $\boldsymbol{T}^2$, and that π is a Riemannian submersion.

Let $H_p (\subset T_p P)$ be the vector space generated by e_1, e_2 at each $p \in P$. Then, H_p defines a connection $\widetilde{\nabla}$ on P, whose curvature is given by $\Theta = -2\pi dx \wedge dy$ (a uniform magnetic field). Furthermore, $\tilde{g}$ turns out to be the Kaluza-Klein metric induced from $\widetilde{\nabla}$. In this situation the spectrum of the Bochner-Laplacian L_m ($m \neq 0$) consists of eigenvalues

$$\lambda_j^{(m)} = 2\pi|m|(2j + 1) \quad (j = 0, 1, 2, \ldots)$$

with multiplicity $|m|$ ([**8**]).

2.2.3. *Harmonic magnetic fields on complex projective spaces* ([**19**]). Consider the vector space $\boldsymbol{C}^{n+1} = \{z = (z_0, z_1, \ldots, z_n)\}$ with the Hermitian inner product $\langle z, z' \rangle := \sum_j z_j \bar{z}'_j$. Put $\langle z, z' \rangle_{\boldsymbol{R}} := \mathrm{Re}\langle z, z' \rangle$, and let

$$S_{[2]}^{2n+1} := \{z \in \boldsymbol{C}^{n+1}; \; \langle z, z \rangle_{\boldsymbol{R}} = 4\},$$

which is a sphere of radius 2 in $\boldsymbol{C}^{n+1} \cong \boldsymbol{R}^{2n+2}$. The group $U(1) = \{\lambda \in \boldsymbol{C}; \; |\lambda| = 1\}$ acts on $S_{[2]}^{2n+1}$ by $z = (z_0, \ldots, z_n) \mapsto \lambda z = (\lambda z_0, \ldots, \lambda z_n)$. Note that this action preserves the Hermitian structure $\langle \cdot, \cdot \rangle$. Thus we have a principal $U(1)$-bundle (called the Hopf bundle):

$$\pi : S_{[2]}^{2n+1} \to \boldsymbol{C}P^n := S_{[2]}^{2n+1}/U(1).$$

The tangent bundle of $S_{[2]}^{2n+1}$ is given by

$$TS_{[2]}^{2n+1} = \{(z, u) \in S_{[2]}^{2n+1} \times \boldsymbol{C}^{n+1}; \; \langle z, u \rangle_{\boldsymbol{R}} = 0\},$$

and the tangent space $T_z S_{[2]}^{2n+1}$ at $z \in S_{[2]}^{2n+1}$ is decomposed as

$$H_z \oplus V_z := \{(z, u); \; \langle z, u \rangle = 0\} \oplus \{(z, icz); \; c \in \boldsymbol{R}\},$$

where two subspaces are orthogonal with respect to $\langle \cdot, \cdot \rangle$, and V_z is tangent to the fiber of the bundle $S_{[2]}^{2n+1}$. This decomposition defines a connection $\widetilde{\nabla}$ on the bundle

$S^{2n+1}_{[2]}$. Take the Riemannian metric g on $\boldsymbol{C}P^n$ such that $\pi_*|_{H_z}$ is an isometry. Then g is a Kähler metric on $\boldsymbol{C}P^n$, called the Fubini-Study metric, whose holomorphic sectional curvature is equal to 1, and the associated fundamental two-form is just the curvature Θ of $\widetilde{\nabla}$. Furthermore, Θ is a harmonic form on $(\boldsymbol{C}P^n, g)$, and $[\Theta/2\pi]$ is a generator of $H^2(\boldsymbol{C}P^n, \boldsymbol{Z}) \cong \boldsymbol{Z}$.

Finally, the spectrum of the Bochner-Laplacian L_m ($m \in \boldsymbol{Z}$) associated to Θ is found to be a set of eigenvalues

$$\lambda_j^{(m)} = \Big(j + \frac{|m|}{2}\Big)\Big(j + \frac{|m|}{2} + n\Big) - \frac{m^2}{4} \quad (j = 0, 1, 2, \ldots).$$

REMARK. In the dynamical systems in Examples 2 and 3 all classical orbits are periodic, and the systems are completely integrable (in the sense of Liouville).

3. Asymptotic behavior of eigenvalues and holonomy of the connection

From now on we consider mainly the first order pseudodifferential operator $\sqrt{L_m}$ and its spectrum $\{\nu_j^{(m)} := (\lambda_j^{(m)})^{1/2}\}$. The operator $\sqrt{L_m}$ is locally defined (modulo a smoothing operator) by

$$\sqrt{L_m}u(x) = (2\pi)^{-n} \int_{\boldsymbol{R}^n} e^{i(x-y)\cdot\xi} p(x,\xi) u(y) dy d\xi \quad (u \in C_0^\infty(U) \cong C_0^\infty(E|_U))$$

with respect to a local section e_m of E_m (see [**29**], for example). Here $p(x,\xi)$ (called the symbol of $\sqrt{L_m}$) is a C^∞ function on $U \times (\boldsymbol{R}^n \backslash 0)$ that admits an asymptotic expansion as $\|\xi\| \to \infty$:

$$p(x,\xi) \sim p_1(x,\xi) + p_0(x,\xi) + \cdots + p_k(x,\xi) + \cdots$$

with $p_k(x,\xi)$ homogeneous of degree k in ξ. In particular,

$$\begin{aligned} p_1(x,\xi) &= \sqrt{H_0(x,\xi)}, \\ p_0(x,\xi) &= -m \sum A^j \xi_j \|\xi\|^{-1} + \frac{i}{2} \sum g^{jk} \Gamma^l_{jk} \xi_l \|\xi\|^{-1} \\ &\quad + \frac{i}{4} \sum \frac{\partial g^{lm}}{\partial x^j} g^{jk} \xi_k \xi_l \xi_m \|\xi\|^{-3}. \end{aligned} \tag{3.1}$$

Here $p_1(x,\xi)$ is called the principal symbol (denoted by $\sigma_P(\sqrt{L_m})$), and is regarded as a function on $T^*U \backslash 0$. Furthermore, $p_1(x,\xi) = \sqrt{H_0(x,\xi)}$ is independent of the choice of the local section e_m, and is well-defined globally on $T^*M \backslash 0$.

Colin de Verdière [**5**] and Chazarain [**4**] obtained beautiful formulas which clarify (under some assumptions) the relations between the spectrum $\{\nu_j^{(0)}\}$ of $\sqrt{\Delta_M}$ $(= \sqrt{L_0})$ and the lengths of closed geodesics on (M, g). Their formulas are generalizations of Poisson's summation formula:

$$\sum_{k \in \boldsymbol{Z}} \exp(ikt) = 2\pi \sum_{k \in \boldsymbol{Z}} \delta(t - 2\pi k)$$

for the circle $\boldsymbol{R}/\boldsymbol{Z}$. Their main tool is the analysis of the kernel of the unitary operator $\exp(-i\sqrt{L_m})$ or the fundamental solution of the wave equation

$$\frac{\partial^2 u}{\partial t^2} + L_m u = 0 \quad \text{i.e.,} \quad \frac{\partial u}{\partial t} + i\sqrt{L_m}u = 0, \tag{3.2}$$

that is treated in the theory of pseudodifferential and Fourier integral operators developed mainly by Hörmander. Later these results were sharpened in a more generalized situation by Duistermaat and Guillemin [**6**]. However, their discussions do not contain the analysis of the lower order terms of L_m associated to the magnetic field.

In this section, on the basis of the theory of pseudodifferential and Fourier integral operators, we will consider the relation between the connection of E (i.e., the lower order terms of L_m) and the (asymptotic) distribution of the spectrum.

3.1. Helton's theorem and its generalization. For the spectrum $\{\nu_j^{(m)}\}$ of $\sqrt{L_m}$, let $\Sigma^{(m)}$ be the set consisting of all the cluster points of the set $\{\nu_p^{(m)} - \nu_q^{(m)};\ p,q \in \boldsymbol{N}\}$. Obviously $\Sigma^{(m)}$ is a closed set of $\boldsymbol{R}$. In the case $m = 0$ Helton [**13**] and Guillemin [**10**] proved the following.

THEOREM (Helton and Guillemin). *The set $\Sigma^{(0)}$ is equal to $\boldsymbol{R}$ or $\{2n\pi/T; n \in \boldsymbol{Z}\}$ for some positive constant T. Moreover, $\Sigma^{(0)} = \{2n\pi/T;\ n \in \boldsymbol{Z}\}$ if and only if all the orbits of the geodesic flow ϕ_t are periodic with the least common period T. Here, the geodesic flow ϕ_t is the Hamiltonian flow on $T^*M\backslash 0$ induced from $p_1(x,\xi) = \sqrt{H_0(x,\xi)}$.*

We give a historical note about this theorem. First, Helton [**13**] showed that

$$\Sigma^{(0)} \neq \boldsymbol{R} \implies \text{all the orbits of the geodesic flow } \phi_t \text{ are periodic.}$$

Next, we note a result of Wadsley [**34**]: If all the orbits of the geodesic flow are closed, then they have a common period. Moreover, if T is the least common period, then almost every orbit is periodic with T as its least period. Such a Riemannian manifold is called a *P_T-manifold* ([**3**]). On the other hand, in the case where all the orbits of the geodesic flow ϕ_t are periodic with period T, Guillemin [**10**] analyzed $\exp(-it\sqrt{\Delta_M})$ in the theory of Fourier integral operators, and showed that for all $\varepsilon > 0$ the spectrum $\{\nu_j^{(0)}; j \in \boldsymbol{N}\}$ without finitely many eigenvalues is contained in

$$\bigcup_{k\in\boldsymbol{N}} \left[\frac{2\pi}{T}(k+\frac{\mu}{4}) - \varepsilon, \frac{2\pi}{T}(k+\frac{\mu}{4}) + \varepsilon\right]$$

(μ being the (constant) Maslov index of the periodic orbits). Combining these results, we have the theorem above.

REMARK. Typical examples of P_T-manifolds are compact symmetric spaces of rank one (the sphere and the projective spaces).

We will consider the structure of $\Sigma^{(m)}$ for $m \neq 0$. For this purpose we make use of the pseudodifferential operators on the principal bundle P associated to those on E_m. Take a local coordinate neighborhood U of M, and let (x,θ) ($x \in U, 0 \le \theta < 2\pi$) be local coordinates on $\overline{U} := \pi^{-1}(U) \cong U \times U(1)$ ($U(1) = \{e^{i\theta}; 0 \le \theta < 2\pi\}$). Consider a pseudodifferential operator $A:\ C^\infty(P) \to C^\infty(P)$ on P locally defined by

$$Au(x,\theta) := \int_{\boldsymbol{R}^n} \left(\sum_{\tau\in\boldsymbol{Z}} e^{ix\cdot\eta} e^{i\theta\tau} a(x,\theta,\eta,\tau)\hat{u}(\eta,\tau)\right) d\eta. \tag{3.3}$$

Here $\hat{u}(\eta,\tau)$ is the Fourier transform of $u = u(x,\theta) \in C_0^\infty(\overline{U})$, and the symbol $a(\bar{x},\bar{\eta})$ ($\bar{x} := (x,\theta), \bar{\eta} := (\eta,\tau)$) is a C^∞ function on $\overline{U} \times \boldsymbol{R}^{n+1}\backslash 0 \cong T^*\overline{U}\backslash 0$ admitting the asymptotic expansion

$$a(\bar{x},\bar{\eta}) \sim \sum_{j=0}^{\infty} a_{k-j}(\bar{x},\bar{\eta}).$$

For $g = e^{i\theta} \in U(1)$, define the operator $T_g : C^\infty(P) \to C^\infty(P)$ by $(T_g f)(p) := f(p\cdot g)\,(p \in P)$. Then, we say a pseudodifferential operator A on P is *ρ_k-equivariant* if $T_g \circ A = \rho_k(g^{-1}) A \circ T_g$ holds for any $g \in U(1)$.

LEMMA. *A ρ_k-equivariant pseudodifferential operator A on P has the following properties.*

(1) $f \in C^\infty_m(P) \implies Af \in C^\infty_{m+k}(P)$.

(2) Let $a(\bar{x},\bar{\eta})$ be the principal symbol of A. Then,

$$a((\bar{x},\bar{\eta})\cdot g) = \rho_k(g^{-1}) a(\bar{x},\bar{\eta}) \quad i.e., \quad a(x,\theta+\theta',\eta,\tau) = e^{-ik\theta'} a(x,\theta,\eta,\tau)$$

for all $g \in U(1)$; namely, $a(\bar{x},\bar{\eta})$ is ρ_k-equivariant.

Using Lemma (1) above and (1.6), we introduce an operator $A^{(m)}_k : C^\infty(E_m) \to C^\infty(E_{m+k})$ associated to a ρ_k-equivariant operator A with $A^{(m)}_k := \psi_{m+k} \circ A \circ \psi_m^{-1}$. Take the local section $s^{(m)}_0$ of the line bundle E_m associated to a local section of P. Then, for $s = f s^{(m)}_0 \in C^\infty_0(E_m|_U)$ $(f \in C^\infty_0(U))$ we have

$$A^{(m)}_k s(x) = \left(\int_{\boldsymbol{R}^n} e^{ix\cdot\eta} a^{(m)}_k(x,\eta) \hat{f}(\eta)\, d\eta \right) s^{(m+k)}_0(x) \quad (x \in U),$$

where the symbol $a^{(m)}_k$ is given by $a^{(m)}_k(x,\eta) = a(x,0,\eta,m)$, with a being the symbol of A.

Now, we will analyze $\Sigma^{(m)}$, intending to generalize Helton's theorem. Put $\tilde{\nu}^{(m)}_j := (\lambda^{(m)}_j + |m|^2/c^2)^{1/2}$. Then it follows from Fact 2.4 that the set of all the cluster points of $\{\tilde{\nu}^{(m)}_p - \tilde{\nu}^{(m)}_q ; p,q \in \boldsymbol{N}\}$ is the same as $\Sigma^{(m)}$. It is easy to see that $\lambda^{(-m)}_j = \lambda^{(m)}_j$, a fact that corresponds to such classical phenomenon as the fact that the orbits of two particles with charge $\pm m$ are the same except for their directions. Hence we have $\Sigma^{(m)} = \Sigma^{(-m)}$.

Let A be a ρ_{2m}-equivariant pseudodifferential operator on P of order zero, and consider $A(t) := \exp(it\sqrt{\Delta_P}) A \exp(-it\sqrt{\Delta_P})$ $(t \in \boldsymbol{R})$. Egorov's theorem asserts that for each $t \in \boldsymbol{R}$, $A(t)$ is a ρ_{2m}-equivariant pseudodifferential operator of order zero whose symbol $a(s)$ is given by $a(s)(\bar{x},\bar{\eta}) = a(\bar{\phi}_t(\bar{x},\bar{\eta}))$. Here, $\bar{\phi}_t(\bar{x},\bar{\eta})$ denotes the orbit of the (homogeneous) geodesic flow on $T^*P\backslash 0$ such that $\bar{\phi}_0(\bar{x},\bar{\eta}) = (\bar{x},\bar{\eta})$. From the operator $A(t)$ we obtain the pseudodifferential operator $A(t)^{(-m)}_{2m} : C^\infty(E_{-m}) \to C^\infty(E_m)$ on line bundles. Furthermore, we consider the operator associated to $f \in C^\infty_0(\boldsymbol{R})$:

$$A_f := \int_{-\infty}^{\infty} \hat{f}(t) A(t)^{(-m)}_{2m}\, dt.$$

The kernel $K(x,y)$ of A_f is given by

$$K(x,y) = \sum_{p,q} f\left(\mu^{(m)}_p - \mu^{(-m)}_q\right) \varphi^{(m)}_p(x) A_{pq} \overline{\varphi^{(-m)}_q}(y), \tag{3.4}$$

where $\{\lambda^{(m)}_j, \varphi^{(m)}_j\}$ is the orthonormal system of the eigenvalues and the associated eigenfunctions of L_m, and $A^{(-m)}_{2m} \varphi^{(-m)}_q = \sum_p A_{pq} \varphi^{(m)}_p$.

Now, we suppose that $\Sigma^{(m)} = \Sigma^{(-m)} \neq \boldsymbol{R}$. Then there exists an open interval I such that the intersection $I \cap \{\nu_p^{(\pm m)} - \nu_q^{(\pm m)};\ p, q \in \boldsymbol{N}\}$ is a finite set. Take a function f such that supp$f \subset I$. Then, $K(x,y)$ is a C^∞ function, because by (3.4) it is a sum of finitely many C^∞ functions. Hence A_f is a compact operator. As a consequence, using Egorov's theorem and a property of the symbol of the pseudodifferential operator of order zero, we obtain

$$\lim_{T\to\infty} \left[\limsup_{|\eta|\to\infty} \sup_{x\in M} \left| \int_{-T}^{T} \hat{f}(t) a(\bar{\phi}_t(x,0,\eta,-m))dt \right| \right] = 0. \tag{3.5}$$

Since each orbit on the unit cotangent bundle: $\bar{\phi}_t(x,0,\eta,-m)/\|(\eta,-m)\|$ converges to $\bar{\phi}_t(x,0,\eta,0)/\|(\eta,0)\|$ when $|\eta| \to \infty$, we get from (3.5)

$$\int_{-\infty}^{\infty} \hat{f}(t) a(\bar{\phi}_t(x,0,\eta,0))dt = 0 \quad (\text{for } \forall(x,\eta) \in T^*M). \tag{3.6}$$

In order for (3.6) to be satisfied for any ρ_{2m}-equivariant function $a(x,\theta,\eta,\tau)$ homogeneous in (η,τ) of order zero, it is necessary that the projection $\phi_t(x,\eta)$ of an orbit $\bar{\phi}_t(x,0,\eta,0)$ to T^*M is periodic. Here, note that $\phi_t(x,\eta)$ is an orbit of the geodesic flow on $T^*M\backslash 0$. Thus, we see that ([**25**])

if $\Sigma^{(m)} \neq \boldsymbol{R}$, then every orbit of the geodesic flow ϕ_t on $T^*M\backslash 0$ is periodic; namely, (M,g) is a P_T-manifold.

Next, suppose (M,g) is a P_T-manifold; namely, $\phi_t(x,\eta)$ is periodic with period T for all (x,η). Then, we have

$$\bar{\phi}_{t+T}(x,0,\eta,0) = \bar{\phi}_t(x,0,\eta,0) \cdot e^{i\pi h} \quad (h = h(x,\eta) \in (-1,1])$$

with $e^{i\pi h}$ being the holonomy of $\widetilde{\nabla}$. Since $a(x,\theta,\eta,\tau)$ is a ρ_{2m}-equivariant, we see that

$$a(\bar{\phi}_t(x,0,\eta,0)) = \sum_{n\in\boldsymbol{Z}} a_n(x,\eta) \exp\Big(i\frac{2\pi n}{T}t\Big) \exp\Big(i\frac{2\pi mh}{T}t\Big).$$

Plugging this into (3.6), we get

$$\int_{-\infty}^{\infty} \sum_n \hat{f}(t) a_n(x,\eta) \exp\Big\{it\Big(\frac{2\pi n}{T} + \frac{2\pi mh}{T}\Big)\Big\}dt$$
$$= \sum_n a_n(x,\eta) f\Big(\frac{2\pi n}{T} + \frac{2\pi mh}{T}\Big) = 0.$$

Since this condition is satisfied for any rapidly decreasing sequence $\{a_n(x,\eta)\}$, we conclude for the function f that

$$f\Big(\frac{2\pi n}{T} + \frac{2\pi mh}{T}\Big) = 0 \quad (\text{for every } n \in \boldsymbol{Z}). \tag{3.7}$$

Let $\Sigma_{\widetilde{\nabla}}$ be the set of the holonomies ($\in U(1)$) of $\widetilde{\nabla}$ along all of the closed geodesics (of length T). It is easy to see that $\Sigma_{\widetilde{\nabla}}$ is a closed subset of $U(1)$ given by $\{e^{it}; -\pi\alpha \le t \le \pi\alpha\}$ or $\{e^{it}; \pi - \pi\alpha \le t \le \pi + \pi\alpha\}$ for some $\alpha \in [0,1]$. As a consequence, from (3.7) we have

$$\Sigma^{(m)} \supset \bigcup_{n\in\boldsymbol{Z}} \Big[\frac{2\pi}{T}(n - m\alpha), \frac{2\pi}{T}(n + m\alpha)\Big]. \tag{3.8}$$

Of course this formula holds in the case $\Sigma^{(m)} = \boldsymbol{R}$. By virtue of (3.8) we have the following.

PROPOSITION 3.1 ([**25**]). *Suppose (M,g) be a P_T-manifold.*
(1) If $\Sigma^{(m)} \neq \boldsymbol{R}$, then $0 \leq \alpha < 1/(2|m|)$.
(2) If $\Sigma^{(m)} = \{2n\pi/T;\ n \in \boldsymbol{Z}\}$ $\forall m \in \boldsymbol{Z}$, then $\Sigma_{\widetilde{\nabla}} = \{1\}$ or $\{-1\}$.

3.2. The analysis for $U(t) = \exp(-it\sqrt{L_m})$ and the asymptotic behavior of eigenvalues. We will analyze the unitary operator $U(t) = \exp(-it\sqrt{L_m})$ ($t \in \boldsymbol{R}$) on E_m induced from $\sqrt{L_m}$ in the theory of Fourier integral operators. The kernel $U(t,x,y)$ of $U(t)$ for sufficiently small $|t|$ can be locally (in (x,y)) given modulo C^∞ functions by a Fourier integral distribution of the form

$$U(t,x,y) = (2\pi)^{-1} \int_{\boldsymbol{R}^n} e^{i\{w(t,x,\xi)-y\cdot\xi\}} K(t,x,\xi)d\xi.$$

Here, by virtue of (3.2), $w(t,x,\xi)$ is homogeneous of degree one in ξ, and governed by the eikonal equation (the Hamilton-Jacobi equation)

$$\partial w/\partial t + p_1(x,w_x) = 0, \quad w(0,x,\xi) = x\cdot\xi, \tag{3.9}$$

and $K(t,x,\xi)$ has an asymptotic expansion

$$K(t,x,\xi) \sim \sum_{j=0}^{\infty} K_{-j}(t,x,\xi),$$

where $K_{-j}(t,x,\xi)$ is homogeneous of degree $-j$ in ξ and is determined inductively by the so-called transport equations. In particular, K_0 is determined by

$$\begin{aligned} &\frac{1}{i}\frac{\partial K_0}{\partial t} + \frac{1}{i}(\partial_\xi p_1)(x,w_x)\cdot\partial_x K_0 + p_0(x,w_x)K_0 \\ &\quad + \frac{1}{2i}\sum_{|\alpha|=2}(\partial_\xi^\alpha p_1)(x,w_x)(\partial_x^\alpha w)K_0 = 0, \end{aligned} \tag{3.10}$$

$$K_0(0,x,\xi) = 1,$$

where p_1 and p_2 are given by (3.1). The characteristic curves of the eikonal equation (3.9) are just the orbits of the (homogeneous) geodesic flow on $T^*M\backslash 0$. Let $(x(t),\xi(t))$ be an orbit of the geodesic flow such that $x(t) = x, \xi(0) = \xi$. Then, the solution w satisfies $w_x(t,x,\xi) = \xi(t)$, $w_\xi(t,x,\xi) = x(0)$. The wave-front set of $U(t)$ is contained in the Lagrangian submanifold $C_t := \{(x,\xi(t);x(0),\xi)\} \subset (T^*M\backslash 0)\times(T^*M\backslash 0)$.

Consider the space $L^2(\Lambda^{1/2}E_m) := L^2(E_m)\otimes\Lambda^{1/2}$ of E_m-valued half-densities on M. Put $\lambda_0 := |\det g(x)|^{1/4}|dx^1\wedge\cdots\wedge dx^n|^{1/2}$, and let each $\lambda \in L^2(\Lambda^{1/2}E_m)$ be written as $\lambda(x) = f(x)\lambda_0(x)$ ($f \in L^2(E_m)$). Hence, we have a one-to-one unitary correspondence: $L^2(\Lambda^{1/2}E_m) \cong L^2(E_m)$; $\lambda \mapsto f$. By virtue of this correspondence the operator P acting on $L^2(E_m)$ can be regarded as the one acting on $L^2(\Lambda^{1/2}E_m)$ defined by $P\lambda := (Pf)\lambda_0$. Hereafter we regard $\sqrt{L_m}$ and $U(t)$ as operators acting on $L^2(\Lambda^{1/2}E_m)$, and consider their symbols.

For the vector field $v := (\partial_\xi p_1)\cdot\partial_x$ and the half-density $|dx|^{1/2}$ on M, the Lie derivative $\mathcal{L}_v(|dx|^{1/2})$ is given by

$$\begin{aligned} \mathcal{L}_v(|dx|^{1/2}) &= \frac{1}{2}\sum_j \frac{\partial}{\partial x^j}\Big(\frac{\partial p_1}{\partial \xi_j}(x,w_x)\Big)|dx|^{1/2} \\ &= \frac{1}{2}\left[\sum_j \frac{\partial^2 p_1}{\partial x^j\partial\xi_j}(x,w_x) + \sum_{|\alpha|=2}(\partial_\xi^\alpha p_1)(x,w_x)(\partial_x^\alpha w)\right]|dx|^{1/2}. \end{aligned}$$

Note that the orbits of the vector field v are geodesics on M. Set $J(t,x,\xi) := \det[\partial x/\partial x(0)] = \det[\partial^2 w/\partial x \partial \xi]^{-1}$, and let d/dt mean the derivative along the orbits of the geodesic flow. Then, we can rewrite equation (3.1) as

$$\frac{1}{i}\frac{d}{dt}\big(|J|^{1/2}K_0\big) + \sigma_{\mathrm{sub}}(\sqrt{L_m})|J|^{1/2}K_0 = 0, \tag{3.11}$$

where

$$\sigma_{\mathrm{sub}}(\sqrt{L_m}) := p_0(x,\xi) - \frac{1}{2i}\sum_j \frac{\partial^2 p_1}{\partial x^j \partial \xi_j}(x,\xi) = -m\sum_j A^j\xi_j/\|\xi\|, \tag{3.12}$$

which is called the *subprincipal symbol* of $\sqrt{L_m}$. The (locally defined) half-density $|J|^{1/2}K_0|dx|^{1/2}$ is called the principal symbol of $U(t)$. In the change of coordinates $(x,\xi) \mapsto (\tilde{x},\tilde{\xi})$, we have $\exp(\pi i\kappa/2)|J|^{1/2}K_0|dx|^{1/2} = |\tilde{J}|^{1/2}\widetilde{K}_0|d\tilde{x}|^{1/2}$ ($\kappa \in \boldsymbol{Z}$), so the principal symbol of $U(t)$ is globally defined as a section of a $\boldsymbol{Z}_4$-bundle called the *Maslov bundle* on the wave-front set C_t.

Suppose (M,g) is a P_T-manifold; namely, every orbit of the geodesic flow is periodic with period T. Take a partition $0 = t_0 < t_1 < \cdots < t_k = T$ with $t_j - t_{j-1}$ sufficiently small. Then, we have a decomposition

$$U(T) = U(t_k,t_{k-1})U(t_{k-1},t_{k-2})\cdots U(t_1)$$

such that the kernel of $U(t_j,t_{j-1})$ is given by a Fourier integral distribution. Since the wave-front set of $U(T)$ is $\{(x,\xi,x,\xi);(x,\xi)\in T^*M\backslash 0\}$, $U(T)$ is a pseudodifferential operator of order zero whose principal symbol is given by

$$\sigma_P\big(U(T)\big)(x,\xi) = e^{-\pi i\mu/2}\Big(\prod_{j=1}^{k}|J^{(j)}|^{1/2}K_0^{(j)}\Big). \tag{3.13}$$

The factor $e^{-\pi i\mu/2}$ is the product of those in the formulas connecting $U(t_{j+1},t_j)$ and $U(t_j,t_{j-1})$ under the change of coordinates, where μ is an integer called the Maslov index of the closed orbits of the geodesic flow.

It follows from (3.12) that the equation (3.11) for $|J^{(j)}|^{1/2}K_0^{(j)}$ along the orbit $(x(t),\xi(t))$ is just the equation of parallel displacement along the geodesic $\gamma := x(t)$ with respect to the connection $\widetilde{\nabla}^{(m)}$ on E_m. Let $Q^{(m)}_{\widetilde{\nabla}}(x,\xi)$ ($\in U(1)$) denote the holonomy of $\widetilde{\nabla}^{(m)}$ along the closed geodesic on M of the initial data $(x,\xi)\in T^*M\backslash 0$. Then, we see from (3.13) that

$$\sigma_P\big(U(T)\big)(x,\xi) = e^{-\pi i\mu/2}Q^{(m)}_{\widetilde{\nabla}}(x,\xi). \tag{3.14}$$

Now, consider the so-called *return operator* $R := e^{\pi i\mu/2}U(T)$, whose eigenvalues are

$$\hat{\nu}_j^{(m)} := \exp\Big\{-2\pi i\Big(\frac{T}{2\pi}\nu_j^{(m)} - \frac{\mu}{4}\Big)\Big\} \quad (j = 1,2,\ldots).$$

Put $\mu_k := \frac{2\pi}{T}(k + \frac{\mu}{4})$ ($k \in \boldsymbol{N}$), and take the interval $I_k := [\mu_k,\mu_{k+1})$. Let us consider the set of eigenvalues $\mathcal{N}_k := \{\hat{\nu}_j^{(m)};\ \nu_j^{(m)} \in I_k\}$ of R, which is a set of finitely many points in the unit circle $S^1 \subset \boldsymbol{C}$. We will see that as $k\to\infty$ the distribution of $\mathcal{N}_k$ in S^1 can be described by the holonomy $Q^{(m)}_{\widetilde{\nabla}}$.

First, suppose $Q^{(m)}_{\widetilde{\nabla}}(x,\xi) = \{1\}$ (or $\{-1\}$) holds for all $(x,\xi)\in S^*M$. Then, $R - I$ (or $R + I$) is a pseudodifferential operator of order -1. Hence, $\sqrt{L_m}(R-I)$

(or $\sqrt{L_m}(R+I)$) is of order zero, and accordingly is a bounded operator. This implies that

$$\max_{\nu\in\mathcal{N}_k} |\nu - 1| \leq \text{Const.}/k \quad (\text{or } \max_{\nu\in\mathcal{N}_k} |\nu + 1| \leq \text{Const.}/k).$$

As a consequence, Proposition 3.1, (2) is sharpened as follows.

THEOREM 3.2. *We have* $\Sigma^{(m)} = \{2n\pi/T;\ n \in \boldsymbol{Z}\}$ *for some* $m \neq 0$ *(then for every* $m \in \boldsymbol{Z}$*) if and only if* (M, g) *is a* P_T*-manifold and the connection* $\widetilde{\nabla}$ *satisfies* $\Sigma_{\widetilde{\nabla}} = \{1\}$ *or* $\{-1\}$.

Next, consider the general cases concerning $\widetilde{\nabla}$. Take the measure $d\omega$ on the cosphere bundle $S^*M := \{(x,\xi);\ \|\xi\| = 1\}$ that is induced from the Liouville measure $(\Omega_M)^n := dx^1 \cdots dx^n d\xi_1 \cdots d\xi_n$ on T^*M. We need the following assumption on E_m.

Assumption. On the vector bundle E_m there exists an (essentially) self-adjoint, elliptic, first order pseudodifferential operator P_0 such that
(i) $\sigma_P(P_0) = \sqrt{H_0}$ $(= \sigma_P(\sqrt{L_m}))$, and
(ii) the spectrum of P_0 without finitely many eigenvalues is equal to

$$\left\{\frac{2\pi}{T}\left(k + \frac{\mu}{4}\right);\, k(\in \boldsymbol{N}) \geq k_0\right\}.$$

Under this assumption we have the following.

THEOREM 3.3 ([**20**]). *For a continuous function* ρ *on* S^1 *we have as* $k \to \infty$ *the asymptotic formula*

$$\sum_{\hat{\nu}\in\mathcal{N}_k} \rho(\hat{\nu}) = \Big[\frac{1}{(2\pi)^n}\int_{S^*M} \rho(Q_{\widetilde{\nabla}}^{(m)}(x,\xi))\, d\omega(x,\xi)\Big] k^{n-1} + o(k^{n-1}).$$

Outline of the proof. Consider the so-called averaged operator of $\sqrt{L_m}$:

$$(\sqrt{L_m})_{\text{av}} := \frac{1}{T}\int_0^T \exp(-itP_0)\sqrt{L_m}\exp(itP_0)dt.$$

By means of Egorov's theorem $(\sqrt{L_m})_{\text{av}}$ is a first order pseudodifferential operator with principal symbol $\sqrt{H_0}$. Let R_{av} be the return operator associated to $(\sqrt{L_m})_{\text{av}}$. Then, $\sigma_P(R_{\text{av}}) = \sigma_P(R)$. This implies that, corresponding to the eigenvalues $\{\nu_{j_s}^{(m)};\, s = 1, \ldots, N_k\}$ of $\sqrt{L_m}$ contained in I_k, there exists a set of eigenvalues $\{\nu_{j_s};\, s = 1, \ldots, N_k\}$ of $(\sqrt{L_m})_{\text{av}}$ such that

$$|\nu_{j_s}^{(m)} - \nu_{j_s}| < \text{Const.}/k.$$

Using this relation, we consider the distribution of eigenvalues $\{\hat{\nu}_j\} \subset S^1$ of R_{av} associated to $\{\nu_j\}$. Since P_0 commutes with $(\sqrt{L_m})_{\text{av}}$, and accordingly with R_{av}, we set, for a C^∞ function ρ on S^1,

$$\hat{e}_\rho(t) := \text{Trace}\Big[\rho(R_{\text{av}})\exp\Big\{-it\Big(P_0 - \frac{\mu}{4}\Big)\Big\}\Big] = \sum_{k=0}^{\infty}\Big(\sum_{j=1}^{N_k}\rho(\hat{\nu}_j)\Big)\exp\Big(-\frac{2\pi k}{T}t\Big),$$

which is the Fourier transform of $e_\rho(\lambda) = \sum_k a_k \delta(\lambda - 2\pi k/T)$ $(a_k := \sum_j \rho(\hat{\nu}_j))$. In this setting we develop arguments similar to those of Hörmander [**14**]—more precisely, we analyze $R(t) := \rho(R_{\text{av}})\exp\{-it(P_0 - \mu/4)\}$ (and its trace) for sufficiently small $|t|$ as a Fourier integral operator, and obtain the asymptotic expansion of a_k

as $k \to \infty$. Here, the first term is given by the trace of the amplitude function (the principal symbol) of $R(0)$, which is equal to $\rho(Q_{\widetilde{\nabla}}^{(m)}(x,\xi))$.

REMARK. Under the above Assumption we can derive (3.8) from Theorem 3.3. We expect that the Assumption is satisfied for any line bundle over a P_T-manifold. In fact, on any line bundle over a compact symmetric space G/K of rank one there exists a connection $\widetilde{\nabla}$ defined by a harmonic two-form ($\in H_D^2(G/K, \boldsymbol{Z})$). Then, it follows from the G-invariance of $\widetilde{\nabla}$ that $Q_{\widetilde{\nabla}}(x,\xi)$ is a constant function. Hence, by virtue of Theorem 3.2 we can construct a required operator P_0 from $\sqrt{L}$ by adding a suitable operator of order zero.

4. Semi-classical aspects of the spectrum (sketches)

In the preceding section we clarified that the spectrum of $\sqrt{L_m}$ is closely related to the geodesics of (M,g) and the holonomies of the connection along geodesics. However, we have not seen there the relationship between the spectrum and the dynamical system $\mathcal{H}_{\text{mag}} = (T^*M, \Omega, H_0)$ of magnetic flow. Unlike the case of geodesic flows, the dynamical system $\mathcal{H}_{\text{mag}}$ is not homogeneous, and so the orbits on each energy level $X_E := H_0^{-1}(E) \subset T^*M\backslash 0$ behave differently, depending on E. We will see that the dynamical properties on X_E are related to the asymptotic properties of the set $\{\nu_j^{(m)} - \sqrt{E}|m|;\ j \in \boldsymbol{N}\}$ as $|m| \to \infty$. Note that the consideration as $|m| \to \infty$ corresponds to the so-called semi-classical analysis for $\hbar \ll 1$.

4.1. Guillemin-Uribe-Zelditch theory. The spectrum of $\sqrt{\Delta_P}$ is

$$\{\tilde{\nu}_j^{(m)} := [\lambda_j^{(m)} + |m|^2/c^2]^{1/2};\ m \in \boldsymbol{Z}, j \in \boldsymbol{N}\}$$

(§2, Fact 2.4). In [**11**] and [**12**], V. Guillemin and A. Uribe studied the singularity of the 2π-periodic distribution

$$\Upsilon(s) := \sum_{m \in \boldsymbol{Z}} \sum_{j=1}^{\infty} \varphi(\tilde{\nu}_j^{(m)} - m\widetilde{E}) e^{ims} \tag{4.1}$$

on $\boldsymbol{R}$, where φ is a rapidly decreasing function on $\boldsymbol{R}$. Each orbit of the Hamiltonian flow defined by the Hamiltonian $\tilde{h}_0 := (\widetilde{H}_0)^{1/2}$ (the principal symbol of $\sqrt{\Delta_P}$) on $T^*P\backslash 0$ is a solution of (1.3) (if we ignore the difference of the parameter). Let $\tilde{\phi}_t$ be the restriction of this flow to the submanifold $W:\ \tilde{h}_0 \equiv \widetilde{E},\ \tau \equiv 1$. Since $\tilde{\phi}_t$ commutes with the $U(1)$-action, we obtain the flow $\phi_t := \tilde{\phi}_t/U(1)$ on $B(\widetilde{E}) := W/U(1)$. The flow ϕ_t is the same (ignoring the difference of the parameter) as the flow on the energy level $H_0 \equiv E := \widetilde{E}^2 - 1/c^2$ of $\mathcal{H}_{\text{mag}} = (T^*M, \Omega, H_0)$. Let $\tilde{\psi}_t$ be the Hamiltonian flow defined by the principal symbol of $-i\partial/\partial\theta$ on P (which commutes with $\sqrt{\Delta_P}$), and consider its orbit $\tilde{\psi}_t(\bar{x}_0, \bar{\eta}_0) = (x_0, \theta_0 e^{it}, \eta_0, 1)$ on W. Then,

THEOREM 4.1. *The singular support of $\Upsilon(s)$ is contained in*

$$\mathcal{T} := \{t_1\widetilde{E} + t_2;\ \exists(\bar{x},\bar{\eta}) \in W \ \text{ such that } \ \tilde{\psi}_{t_2} \circ \tilde{\phi}_{t_1}(\bar{x},\bar{\eta}) = (\bar{x},\bar{\eta})\}. \tag{4.2}$$

In (4.2) t_1 must be a period of a periodic orbit $\tilde{\gamma}$ of ϕ_t, i.e., $\phi_{t_1}(x,\eta) = (x,\eta) \in B(\widetilde{E}) \cong X_E$. Furthermore, we have $\tilde{\phi}_{t_1}(\bar{x},\bar{\eta}) = (\bar{x},\bar{\eta})\cdot e^{-it_2}$. Let γ be the projection of $\tilde{\gamma}$ on M, and let T_γ denote the period of γ as a (projected) orbit of $\mathcal{H}_{\text{mag}}$. Then, we have the relations $t_1 = 2\widetilde{E}T_\gamma$ and $e^{-it_2} = Q_{\widetilde{\nabla}}(\gamma)e^{2iT_\gamma/c^2}$, with $Q_{\widetilde{\nabla}}(\gamma)$ being the holonomy of $\widetilde{\nabla}$ (cf. (1.3)). Combining the relations above and denoting the set of all periodic orbits on X_E of $\mathcal{H}_{\mathrm{mag}}$ by Γ_E, we have the following.

COROLLARY. *Suppose $s \in \boldsymbol{R}$ belongs to the singular support of $\Upsilon(s)$. Then,*

$$e^{is} \in \{(Q_{\widetilde{\nabla}}(\gamma))^{-1} e^{2iET_\gamma};\ \gamma \in \Gamma_E\}.$$

Furthermore, under a certain assumption on the dynamical system, Guillemin and Uribe derived an asymptotic formula as $m \to \infty$:

$$\sum_{j=1}^{\infty} \varphi(\tilde{\nu}_j^{(m)} - m\widetilde{E}) \sim \sum_{k=0}^{\infty} a_k(\hat{\varphi}, m) m^{d-k},$$

where $d = \frac{1}{2}\dim X_E - 1 = n - \frac{3}{2}$ and the coefficients $a_k(\hat{\varphi}, m)$ have information about the periodic orbits on X_E (their periods, the associated Poincaré maps, the holonomies, the Maslov index, and so on) (see [**11**] and [**12**] for details).

On the other hand, on the basis of the theory of Guillemin and Uribe, S. Zelditch [**37**] obtained the following result, which asserts that the eigenfunctions become uniformly distributed. that For $r > 0$ put $N_m(\widetilde{E}, r) := \{j \in \boldsymbol{N};\ |\tilde{\nu}_j^{(m)} - m\widetilde{E}| \le r\}$. Take the measure $d\omega_B$ on $B(\widetilde{E})$ naturally induced from the Liouville measure on T^*P. Then,

THEOREM 4.2. *Suppose the set of periodic points of the flow ϕ_t on $B(\widetilde{E})$ has measure zero. Then, for any $U(1)$-invariant pseudodifferential operator A on P of order zero, we have*

$$\sum_{j \in N_m(\widetilde{E}, r)} (A\varphi_j^{(m)} | \varphi_j^{(m)}) = 2r\left(\frac{m}{2\pi}\right)^d \int_{B(\widetilde{E})} \sigma_P(A) d\omega_B + o(m^d) \quad (m \to \infty),$$

where $\{\varphi_j^{(m)}\}$ is an orthonormal set of eigenfunctions of $\sqrt{\Delta_P}$, $(\cdot|\cdot)$ denotes the inner product of $C^\infty(P)$, and the principal symbol $\sigma_P(A)$ of A is regarded as a function on $B(\widetilde{E})$.

Later, motivated by this result, various studies on "ergodicity" or "quantum chaos" for quantum dynamical systems in gauge fields have been developed (see [**28**], [**37**], [**31**], [**32**], for example).

4.2. Quantization condition and the spectrum. The so-called Bohr-Sommerfeld quantization condition plays a important role in semi-classical analysis for completely integrable systems. We will here discuss a quantization condition for the dynamical system $\mathcal{H}_{\mathrm{mag}}$, and clarify the relationship between the semi-classical energy levels (satisfying the quantization condition) and the quantum energy levels: $\{\lambda_j^{(m)}\}$ or $\{\nu_j^{(m)}\}$ (see [**23**], [**24**]).

Quantization condition for magnetic flow. Let Λ be a Lagrangian submanifold of (T^*M, Ω). Put

$$\Lambda_P := (\Psi_1 \circ \pi_1)^{-1}(\Lambda) \subset J^{-1}(1) \subset T^*P$$

(see Figure 1). Then, Λ_P is a Lagrangian submanifold of (T^*P, Ω_P). The standard symplectic form Ω_P on T^*P is given as $\Omega_P = d\omega_P$ by the canonical one-form ω_P. Let $m_{\Lambda_P} \in H^1(\Lambda_P, \boldsymbol{Z})$ be the Maslov class of the Lagrangian submanifold Λ_P.

For a Lagrangian submanifold $\Lambda \subset (T^*M, \Omega)$, we consider the following condition:

$$\text{(Q)} \qquad \frac{1}{2\pi}\int_c \omega_P - \frac{1}{4} m_{\Lambda_P}(c) \in \boldsymbol{Z} \quad \text{for any closed curve } c \text{ on } \Lambda_P,$$

which we call *Maslov's quantization condition* for Λ ([**36**]).

REMARK. If the magnetic field Θ is exact, i.e., $\Theta = d\theta$, then we have $\Omega = d\omega$ with $\omega = \omega_M + \pi_M^*\theta$, and the condition (Q) is equivalent to

$$(Q_M) \qquad \frac{1}{2\pi}\int_c \omega - \frac{1}{4} m_{\Lambda}(c) \in \boldsymbol{Z} \quad \text{for any closed curve } c \text{ on } \Lambda.$$

In this notation we have the following theorem, which is a magnetic version of Weinstein's result [**35**] (see also [**33**, Ch. XII, §4]).

THEOREM 4.3 ([**23**]). *Suppose there exists a Lagrangian submanifold Λ of (T^*M, Ω) which satisfies*

(i) $H_0 \equiv E$ on Λ,

(ii) Λ is invariant under the magnetic flow ϕ_t, and moreover there exists a non-zero half-density on Λ which is invariant under the restricted flow $\phi_t|_\Lambda$, and

(iii) Λ satisfies the quantization condition (Q).

Then, there is a sequence $\{\lambda_{j_k}^{(dk+1)}\}_{k=0}^{\infty}$ of eigenvalues of L_{dk+1}'s such that

(4.3)

$$|\lambda_{j_k}^{(dk+1)} - (dk+1)^2 E| < \text{Const.}, \quad \textit{i.e.} \quad |\nu_{j_k}^{(dk+1)} - (dk+1)\sqrt{E}| < \text{Const.}/(dk+1),$$

where d is the smallest element of the set $\{1, 2, 4\}$ for which $dm_{\Lambda_P}(c) \equiv 0 \pmod 4$ for any closed curve c.

Consider the Schrödinger operator $\hat{H}_m$ given by (1.7), and note that the eigenvalue problem $\hat{H}_m\psi = E\psi$ corresponds to $L_m\psi = Em^2\psi$. Put $m_k = dk + 1 = 1/\hbar$. Then, $\lambda(\hbar) := \lambda_{j_k}^{(m_k)}/m_k^2$ is an eigenvalue of $\hat{H}_m$, and (4.3) implies

$$|\lambda(\hbar) - E| < Rm_k^{-2} = R\hbar^2 \quad (R \text{ a constant}).$$

Thus, we can understand Theorem 4.3 as saying that if E is a semi-classical energy (i.e., the energy level:$H_0 \equiv E$ is a Lagrangian submanifold satisfying the quantization condition), then E gives an approximate quantum energy level of order $\hbar^2$.

REMARK. Suppose (T^*M, Ω, H_0) is completely integrable, i.e., there exist commutative first integrals $f_1 = H_0, f_2, \ldots, f_n$. Then each torus

$$\Lambda_C := \{p \in T^*M; f_i(p) = C_i \ (0 \le i \le n)\}$$

is a Lagrangian submanifold and automatically satisfies conditions (i) and (ii) in Theorem 4.3.

Concluding comments

In this article we have discussed classical and quantum mechanics in the magnetic field on a manifold by focusing on their Hamiltonian structure and the associated quantum objects (the pseudodifferential and Fourier integral operators). We remark that the topics and the techniques treated here are rather restricted. The spectral geometry in a magnetic field is formulated on the principal $U(1)$-bundle. If we consider generally a compact Lie group as the structure group of the bundle, we can develop *spectral geometry in a gauge field.* In fact, the preceding results by Guillemin, Uribe, and Zelditch have been obtained in such formulation.

ACKNOWLEDGMENT. The author would like to express his gratitude to the referee for carefully reading the manuscript and offering valuable comments.

References

[1] R. Abraham and J. Marsden, *Foundations of mechanics* (2nd edition), Benjamin Cummings, Reading, Mass., 1978.

[2] M. Berger, P. Gauduchon, and E. Mazet, *Le spectre d'une variété riemannienne*, Lect. Notes in Math., **194**, Springer-Verlag, 1971.

[3] A. Besse, *Manifolds all of whose geodesics are closed*, Springer, Berlin, 1978.

[4] J. Chazarain, Formule de Poisson pour les variétés riemanniennes, Invent. Math. **24**(1974), 65-82.

[5] Y. Colin de Verdière, Spectre du laplacien et longueurs des géodésiques périodiques. I, II, Compositio Math. **27**(1973), 83-106, 159-184.

[6] J.J. Duistermaat and V.W. Guillemin, The spectrum of positive elliptic operators and periodic bicharacteristics, Invent. Math. **29**(1975), 39-79.

[7] P.B. Gilkey, The spectral geometry of a Riemannian manifold, J. Diff. Geom., **10**(1975), 601-618.

[8] C. Gordon and E. Wilson, The spectrum of the Laplacian on Riemannian Heisenberg manifolds, Michigan Math. J., **33**(1986), 253-271.

[9] C.G. Gordon and E.N. Wilson, Isospectral deformations of compact solvmanifolds, J. Diff. Geom., **19**(1984), 241-256.

[10] V. Guillemin, Lectures on spectral theory of elliptic operators, Duke Math. J., **44**(1977), 485-517.

[11] V. Guillemin and A. Uribe, Circular symmetry and the trace formula, Invent. Math., **96**(1989), 385-423.

[12] V. Guillemin and A. Uribe, Reduction and the trace formula, J. Diff. Geom., **32**(1991), 315-347.

[13] W. Helton, An operator algebra approach to partial differential equations: Propagation of singularities and spectral theory, Indiana Univ. Math. J., **26**(1977), 997-1018.

[14] L. Hörmander, The spectral function of an elliptic operator, Acta Math. **121**(1968), 193-218.

[15] L. Hörmander, *The Analysis of Linear Partial Differential Operators*. III, IV, Springer-Verlag, 1985.

[16] S. Kobayashi, *Differential Geometry of Connections and Gauge Theory* (Japanese), Shokabo, Tokyo, 1989.

[17] S. Kobayashi and K. Nomizu, *Foundations of Differential Geometry*. Vol. I, John Wiley, New York, 1963.

[18] B. Kostant, Quantization and unitary representations, *Lectures in Modern Analysis and Applications*. III, Springer Lect. Notes in Math. **170**(1970), 87-208.

[19] R. Kuwabara, Spectrum of the Schrödinger operator on a line bundle over complex projective spaces, Tôhoku Math. J., **40**(1988), 199-211.

[20] R. Kuwabara, Spectrum of the Laplacian on vector bundles over $C_{2\pi}$-manifolds, J. Diff. Geom., **27**(1988), 241-258.

[21] R. Kuwabara, Isospectral connections on line bundles, Math. Z., **204**(1990), 465-473.

[22] R. Kuwabara, Spectra and geodesic flows on nilmanifolds: Reductions of Hamiltonian systems and differential operators, J. Math. Soc. Japan, **46**(1994), 119-137.

[23] R. Kuwabara, On Maslov's quantization condition for mechanics in a magnetic field, J. Math. Tokushima Univ., **33**(1999), 33-54.

[24] R. Kuwabara, Classical orbits for dynamical systems and quantum energy distribution (Japanese), Sūrikaisekikenkyūsho Kōkyūroku, **1119**(1999), 26-34.

[25] R. Kuwabara, Difference spectrum of the Schrödinger operator in a magnetic field, Math. Z., **233**(2000), 579-599.

[26] J. Marsden and A. Weinstein, Reduction of symplectic manifolds with symmetry, Rep. Math. Phys., **5**(1974), 121-130.

[27] H. McKean and I.M. Singer, Curvature and the eigenvalues of the Laplacian, J. Diff. Geom., **1**(1967), 43-69.

[28] R. Schrader and M. Taylor, Semi-classical asymptotics, gauge fields and quantum chaos, J. Funct. Anal. **83**(1989), 258-316.

[29] R. Seeley, Complex powers of an elliptic operator, *Singular Integrals*, Proc. Symp. Pure Math. **10**(1967), 288-307.

[30] T. Sunada, *Fundamental Groups and Laplacian* (Japanese), Kinokuniya, Tokyo, 1988.

[31] T. Tate, Quantum ergodicity at a finite energy level, J. Math. Soc. Japan **51**(1999), 867-885.

[32] T. Tate, Asymptotic behavior of eigenfunctions and eigenvalues for ergodic and periodic systems, Thesis, Tohoku Math. Pub., No.12(1999).

[33] F. Treves, *Introduction to Pseudodifferential and Fourier Integral Operators*, Vol.2, Plenum Press, New York,1980.

[34] A.W. Wadsley, Geodesic foliations by circles, J. Diff. Geom., **10**(1975), 541-549.

[35] A. Weinstein, On Maslov's quantization condition, *Fourier Integral Operators and Partial Differential Equations*, Springer Lect. Notes in Math. **459**(1974), 341-372.

[36] A. Yoshioka, The quasi-classical calculation of eigenvalues for the Bochner-Laplacian on a line bundle, in "*Geometry of Manifolds (Matsumoto, 1988)*", Academic Press, Boston, Mass., 1989, pp. 39-56.

[37] S. Zelditch, On a "quantum chaos" theorem of R. Schrader and M. Taylor, J. Funct. Anal. **109**(1992), 1-21.

DEPARTMENT OF MATHEMATICAL SCIENCES, UNIVERSITY OF TOKUSHIMA, MINAMI-JOSANJIMA, TOKUSHIMA 770-8502, JAPAN

Translated by RUISHI KUWABARA

Amer. Math. Soc. Transl.
(2) Vol. **211**, 2003

Symbolic Dynamics and C^*-algebras

Kengo Matsumoto

0. Introduction

We will deal with two different types of objects. One is discrete dynamical systems, consisting of the shifts on bi-infinite sequences of finite symbols, whose theory is mainly part of topological dynamical systems. The other one is operator algebras, consisting of infinite dimensional algebras of mutually noncommutative operators, whose theory is mainly part of functional analysis. The two objects look very different. In this paper, we will introduce an interesting relationship between them.

The theory of operator algebras since its birth has had influence on many areas of mathematics and mathematical physics: quantum mechanics, representation theory of groups, quantum statistical mechanics, ergodic theory, and also noncommutative differential geometry, quantum groups, knot invariants arising from the Jones index theory of subfactors, conformal field theory, quantum field theory, and recently Voiculescu's free pobability theory, etc. Especially, in connection with ergodic theory, there is an outstanding result from the 1980s by Connes and Krieger; that is, the isomorphism classes of AFD von Neumann algebras of type III exactly correspond to the orbit equivalence classes of nonsingular ergodic transformations on measure spaces. The C^*-algebra setting of this result has been studied by Giordano, Putnam, and Skau in the 90s. They studied minimal homeomorphisms on Cantor sets and their C^*-crossed products ([GPS], [HPS]). J. Tomiyama studied a more general setting for the interplay between topological dynamical systems and C^*-crossed products ([To], [To2]).

Cuntz and Krieger [CK] in 1980 presented many examples of purely infinite simple C^*-algebras associated to Markov chains of symbolic dynamics. These examples have played a fundamental rôle in the recent classification theory of nuclear simple C^*-algebras. Furthermore, the structure of these C^*-algebras gave rise to a topological conjugacy invariant of topological Markov shifts, called the dimension group defined by Krieger, through K-theory for C^*-algebras ([Kr], [Kr2], cf. [Ef]). Cuntz and Krieger also observed a close relationship between the theory of topological Markov shifts and the theory of C^*-algebras, such as the fact that

2000 *Mathematics Subject Classification.* Primary 46L35; Secondary 37B10.
This article originally appeared in Japanese in Sūgaku **53** (3) (2001), 259–278

the Ext-groups of these C^*-algebras coincide with the Bowen-Franks groups of the Markov shifts, which are invariants for flow equivalence.

In this article, after briefly reviewing the Cuntz-Krieger results, we will survey recent results by the author for symbolic dynamics generalizing them. They are studies on relationships between invariants, presentations of general symbolic dynamics, and C^*-algebras.

Symbolic dynamical systems, often called subshifts, form a class of topological dynamical systems, consisting of homeomorphims (or continuous maps) on compact Hausdorff spaces, defined as follows. For a given finite set, which is called an alphabet and whose members are called symbols, and a family of forbidden words of the symbols, the subshift is defined to be a discrete dynamical system of the one-step shift on the set of all two-sided (or one-sided) infinite sequences of symbols that do not contain any forbidden words in any place. An origin of symbolic dynamics is due to the following idea ([Had]): For a given topological dynamical system, let us decompose the space into a finite disjoint union and assign symbols to the partitions. Then the orbit of each point appears as an infinite sequence of the symbols. The transformation of the dynamics corresponds to the shift on the sequences of the symbols. Let us consider a function f on $[0,1]$ defined by $f(x) = 2x \pmod{\mathbb{Z}}$, for $x \in [0,1)$, $f(1) = 1$. Put $\Sigma = \{0,1\}$ and define the shift σ on the space $\Sigma^{\mathbb{N}}$ by $\sigma((a_i)_{i\in\mathbb{N}}) = (a_{i+1})_{i\in\mathbb{N}}$. The dynamics $([0,1], f)$ may be described as the symbolic dynamics $(\Sigma^{\mathbb{N}}, \sigma)$ through the correspondence between a point of $[0,1]$ and the coefficients of its 2-adic expansion realized as an element of $\Sigma^{\mathbb{N}}$. In this case we assign 0 and 1 to the intervals $[0,1/2)$ and $[1/2,1]$, respectively. Its fobidden words are empty. Although the definition of symbolic dynamical system is easy to describe as above, there are not so many mathematicians doing research in the theory of symbolic dynamical systems itself. The theory of symbolic dynamical system, however, gives fundamental tools to study fractal geometry, one-dimensional dynamics in ergodic theory and topological dynamics, hyperbolic systems like Anosov diffeomorphisms in smooth dynamical systems, and so on. As seen in the above definition, it has close relationships to coding theory, language theory, and information theory. It is applied to encode and decode data on floppy disks and personal computers in our life (cf.[LM]). There is an explanation of how audio disks make certain subshifts in recoding Beethoven's symphnies ([Mo]). There are basic classes of subshifts, the class of shifts of finite type, written as SFT, that contains topological Markov shifts, and the class of sofic shifts, often called sofic systems. Subshifts of these classes are naturally defined by finite directed graphs and finite directed labeled graphs respectively. They are related to graph theory and automata theory. The symbolic dynamics arising from β-adic expansions of real numbers, called the β-shifts for $1 < \beta \in \mathbb{R}$, are interesting from the number theoretical viewpoint ([Pa], [Ré], cf. [IT]). As seen in a recent connection with knot invariants and topological quantum field theory ([Gil], [SW]), the theory of symbolic dynamics intersects with many branches of mathematics and mathematical sciences. Recent textbooks of symbolic dynamics are Lind and Marcus's book [LM] and Kitchen's book [Kit]. As recent textbooks on operator algebras, we mention [Da] and [EK]. The textbook [IH] on operator algebras is written from the viewpoint of ergodic theory. Also Tomiyama's books [To] and [To2] are focused on the interplay between topological dynamics and C^*-crossed products.

1. Topological Markov shifts and Cuntz-Krieger algebras

We fix an arbitrary finite set $\Sigma = \{1, 2, \dots, n\}, n > 1$. Their infinite product space $\Sigma^{\mathbb{Z}} = \prod_{\mathbb{Z}} \Sigma$ is a compact Hausdorff space by the infinite product topology. The shift σ defined by $\sigma((x_k)_{k\in\mathbb{Z}}) = (x_{k+1})_{k\in\mathbb{Z}}$ is a homeomorphism on $\Sigma^{\mathbb{Z}}$. Let $A = [A(i,j)]_{i,j=1,2,\dots,n}$ be an $n \times n$ matrix with entries $A(i,j)$ in $\{0,1\}$. Such a matrix is called a $\{0,1\}$-matrix. We set

$$\Lambda_A = \{(x_k)_{k\in\mathbb{Z}} \in \Sigma^{\mathbb{Z}} \mid A(x_k, x_{k+1}) = 1, k \in \mathbb{Z}\},$$

which is a closed subset of $\Sigma^{\mathbb{Z}}$ invariant under σ (namely, $\sigma(\Lambda_A) = \Lambda_A$). The topological dynamical system (Λ_A, σ) is called the topological Markov shift defined by A. It is written as Λ_A for short. In the theory of topological Markov shifts, the problem of classifying topological Markov shifts under topological conjugacy seems to be most important and interesting. For this problem, R. F. Williams [Wi] in 1973 gave the following answer in terms of matrices.

THEOREM 1.1 (R. F. WILLIAMS [WI]). *Two topological Markov shifts Λ_A and Λ_B are topologically conjugate if and only if the matrices A and B are strong shift equivalent.*

Williams defined two matrices A and B to be strong shift equivalent if there exist nonnegative rectangular matrices $R_1, S_1, R_2, S_2, \dots, R_m, S_m$ such that

$$A = R_1 S_1, \quad S_1 R_1 = R_2 S_2, \quad S_2 R_2 = R_3 S_3, \quad \dots, \quad S_m R_m = B.$$

Although Williams's theorem looks like a complete answer for the conjugacy problem of topological Markov shifts, the conjugacy problem is not finished yet, because there has not been found a general algorithm to decide whether or not two given matrices A and B are strong shift equivalent. The theorem says that the conjugacy problem may be replaced by another hard problem. Williams then introduced the following weaker equivalence relation of matrices. Two matrices A and B are said to be shift equivalent if there exist rectangular matrices R, S and a natural number l such that

$$AR = RB, \quad BS = SA, \quad A^l = RS, \quad B^l = SR.$$

An important notion for studying shift equivalence has been introduced by W. Krieger. It is called the dimension group. For a matrix A, define the inductive limits:

$$\Delta_A = \lim \mathbb{Z}^n \xrightarrow{A^t} \mathbb{Z}^n \xrightarrow{A^t} \cdots, \qquad \Delta_A^+ = \lim \mathbb{Z}_+^n \xrightarrow{A^t} \mathbb{Z}_+^n \xrightarrow{A^t} \cdots,$$

where A^t denotes the transpose of A and $\mathbb{Z}_+^n$ is the n-times product of $\mathbb{Z}_+ = \{0, 1, 2, \dots, \}$. The matrix A^t acts on the ordered abelian group (Δ_A, Δ_A^+) by left-multiplication, denoted by δ_A. The triple $(\Delta_A, \Delta_A^+, \delta_A)$ is called the dimension triple for A or the dimension group for A. Krieger has proved in [Kr2], [Kr3] that the dimension group is a complete invariant for shift equivalence. Strong shift equivalence immediately implies shift equivalence. The converse implication has been a long-standing (2 decades) open problem. Kim and Roush have found an example of a pair of shift equivalent irreducible matrices that are not strong shift equivalent ([KimR]).

The classification of topological Markov shifts under flow equivalence is also important and interesting. For a topological Markov shift (Λ_A, σ), the space F_A denotes the quotient space of the product $\Lambda_A \times \mathbb{R}$ under the equivalence relation

$(x, s+1) \sim (\sigma(x), s)$. Define a flow on it by $\phi_t^A([(x,s)]) = [(x, s+t)], t \in \mathbb{R}$. The flow space (F_A, ϕ^A) is called the suspension flow of (Λ_A, σ). Two topological Markov shifts $(\Lambda_A, \sigma), (\Lambda_B, \sigma)$ are said to be flow equivalent if there exists a homeomorphism from F_A to F_B that maps orbits of ϕ^A to orbits of ϕ^B in an orientation-preserving way. For a topological Markov shift (Λ_A, σ), Parry and Sullivan [PS] proved that $\det(\mathrm{Id} - A)$ is an invariant for flow equivalence. Bowen and Franks [BF] proved that $\mathbb{Z}^n/(\mathrm{Id} - A)\mathbb{Z}^n$ is also an invariant for flow equivalence. The latter abelian group is callled the Bowen-Franks group for A and written as $BF(A)$. Franks proved the converse, that is, $\det(\mathrm{Id} - A)$ and $BF(A)$ are a complete set of invariants for flow equivalence of topological Markov shifts ([Fr]). Parry [Pa2] has proved that for an irreducible matrix A, the topological entropy of (Λ_A, σ) is $\log r_A$, where r_A is the Perron-Frobenius eigenvalue of A.

Let us go to operator algebras. We will briefly review K-theory for C^*-algebras (see [Bl]). A complex vector space with product structure and $*$-operation satisfying $(cx + dy)^* = \bar{c}x^* + \bar{d}y^*, (xy)^* = y^*x^*, (x^*)^* = x$ is called a $*$-algebra. A C^*-algebra is a complete normed $*$-algebra satisfying $\|xy\| \le \|x\|\|y\|, \|x^*x\| = \|x\|^2$. A C^*-algebra may be realized as a C^*-subalgebra of the C^*-algebra of all bounded linear operators on a Hilbert space. Hence a C^*-algebra is in general noncommutative and infinite dimensional. Let $\mathcal{A}$ be a unital C^*-algebra. Consider the tensor product $\mathcal{A} \otimes \mathcal{K}$ between $\mathcal{A}$ and the C^*-algebra $\mathcal{K}$ of all compact operators on a separable infinite dimensional Hilbert space. Two projections p, q in $\mathcal{A} \otimes \mathcal{K}$ are said to be equivalent if there exists $u \in \mathcal{A} \otimes \mathcal{K}$ such that $p = u^*u, q = uu^*$; we write $p \sim q$. The relation $\sim$ in the set of all projections of $\mathcal{A} \otimes \mathcal{K}$ is an equivalence relation. We denote by $[p]$ the equivalence class of p. Define $[p] + [q] = [p' + q']$, where p', q' are mutually orthogonal projections such that $p' \sim p, q' \sim q$. The equivalence classes of the projections of $\mathcal{A} \otimes \mathcal{K}$ become an abelian semigroup. The K_0-group of $\mathcal{A}$ is then defined by the enveloping abelian group of the abelian semigroup in the sense of Grothendieck. It is denoted by $K_0(\mathcal{A})$. It is invariant under the isomorphism class of $\mathcal{A}$ and also invariant under taking the tensor product with $\mathcal{K}$. The other K-group, written as $K_1(\mathcal{A})$ and called the K_1-group of $\mathcal{A}$, is defined by the quotient group of the unitary group of the unitization of the algebra $\mathcal{A} \otimes \mathcal{K}$ by its connected component of the unit.

For a $\{0,1\}$-matrix $A = [A(i,j)]_{i,j=1,2,\ldots,n}$, Cuntz and Krieger [CK] in 1980 introduced a C^*-algebra generated by partial isometries $S_1, S_2, \ldots, S_n$ satisfying the relations

$$(1.1) \qquad \sum_{j=1}^{n} S_j S_j^* = 1, \qquad S_i^* S_i = \sum_{j=1}^{n} A(i,j) S_j S_j^*, \quad i = 1, 2, \ldots, n.$$

This C^*-algebra is called the Cuntz-Krieger algebra defined by A and written as $\mathcal{O}_A$. If all its entries $A(i,j)$ are 1, the associated topological Markov shift is the full n-shift and the second relation of (1.1) becomes $S_i^* S_i = 1$. The associated algebra is generated by n isometries with mutually orthogonal ranges, that is, the Cuntz-algebra $\mathcal{O}_n$ ([C]). We put

$$X_A = \{(x_k)_{k \in \mathbb{N}} \mid A(x_k, x_{k+1}) = 1, k \in \mathbb{N}\},$$

the set of all right one-sided sequences of Λ_A. The topological dynamical system (X_A, σ) is said to be a right one-sided topological Markov shift. Cuntz and Krieger said that the matrix A saisfies condition (I) if X_A is homeomorphic to the Cantor discontinuum. This is a mild condition, because irreducible matrices but not permutations satisfy condition (I). Aperiodic matrices also satisfy condition (I). Assume that a matrix A satisfies condition (I). Cuntz and Krieger first proved that the C^*-algebra $\mathcal{O}_A$ is uniquely determined by the operator relations (1.1) up to isomorphism ([CK]). It is a key in analyzing the structure of the algebra $\mathcal{O}_A$ that the correspondence on the generators

$$\alpha_t(S_i) = tS_i, \qquad i = 1, 2, \ldots, n, \quad t \in \mathbb{T} = \{t \in \mathbb{C} \mid |t| = 1\},$$

gives rise to a one-parameter automorphism group on $\mathcal{O}_A$. It is called the gauge action, and is often written as α^A. Cuntz and Krieger proved the following theorem.

THEOREM 1.2 (CUNTZ-KRIEGER [CK]). *If A is irreducible, the C^*-algebra $\mathcal{O}_A$ is simple (no nontrivial ideal). In particular, if A is aperiodic, it is simple and purely infinite.*

The algebra $\mathcal{O}_A$ is nuclear, that is, for any C^*-algebra $\mathcal{B}$, the tensor product $\mathcal{O}_A \otimes \mathcal{B}$ has a unique C^*-norm. The class of nuclear C^*-algebras contains both the abelian C^*-algebras and the full matrix algebras. The class seems to be considered less difficult to treat than the class of nonnuclear C^*-algebras. A C^*-algebra $\mathcal{A}$ is said to be purely infinite if any hereditary subalgebra of $\mathcal{A}$ contains an infinite projection, that is, any hereditary subalgebra of $\mathcal{A}$ does not have a finite trace, such as a trace on a matrix algebra. Examples of nuclear, simple, purely infinite C^*-algebras had not been found before Theorem 1.2, except for Cuntz algebras. Thanks to Cuntz and Krieger's work, operator algebraists have many examples of nuclear, simple, purely infinite C^*-algebras. The K-groups of the Cuntz-Krieger algebras were computed as

$$K_0(\mathcal{O}_A) = \mathbb{Z}^n/(\mathrm{Id} - A^t)\mathbb{Z}^n, \qquad K_1(\mathcal{O}_A) = \mathrm{Ker}(\mathrm{Id} - A^t)$$

by Cuntz [C2]. Any finitely generated abelian group is realized as the K_0-group of a Cuntz-Krieger algebra, and the K_1-group is its torsion-free part ([Rø]). It is also important that the fixed point algebra of the gauge action

$${\mathcal{O}_A}^{\alpha} = \{X \in \mathcal{O}_A \mid \alpha_t(X) = X, t \in \mathbb{T}\}$$

is the AF-algebra defined by the embedding through the matrix A^t. We denote the algebra by $\mathcal{F}_A$. AF-algebras are C^*-algebras defined by inductive limits of finite dimensional algebras, and their structures are completely determined by their embedding rules ([Bra], cf. [Ef]).

The following theorem, proved by Cuntz and Krieger, plays a key rôle in studying topological Markov shifts as topological dynamics by using C^*-algebras.

THEOREM 1.3 (CUNTZ-KRIEGER [CK]). *If two topological Markov shifts (Λ_A, σ) and (Λ_B, σ) are topologically conjugate, the stabilized C^*-algebras $\mathcal{O}_A \otimes \mathcal{K}$ and $\mathcal{O}_B \otimes \mathcal{K}$ with the gauge actions are isomorphic. That is, there exists an isomorphism ψ from $\mathcal{O}_A \otimes \mathcal{K}$ onto $\mathcal{O}_B \otimes \mathcal{K}$ such that $\psi \circ (\alpha_t^A \otimes \mathrm{id}) = (\alpha_t^B \otimes \mathrm{id}) \circ \psi, t \in \mathbb{T}$.*

This theorem states that the C^*-dynamical system $(\mathcal{O}_A \otimes \mathcal{K}, \alpha^A \otimes \mathrm{id}, \mathbb{T})$ is a topological conjugacy invariant of the topological Markov shift (Λ_A, σ). Therefore one has

COROLLARY 1.4. *The following abelian groups for $\mathcal{O}_A$ are invarints for topological conjugacy of topological Markov shift (Λ_A, σ):*

$$K_0(\mathcal{O}_A), \quad K_1(\mathcal{O}_A), \quad \mathrm{Ext}^0(\mathcal{O}_A), \quad \mathrm{Ext}^1(\mathcal{O}_A).$$

The groups $\mathrm{Ext}^*(\mathcal{O}_A), * = 0, 1$, are often called the K-homology groups of $\mathcal{O}_A$, and correpond to the dual objects of $K_*(\mathcal{O}_A)$. They are defined as $\mathrm{Ext}^1(\mathcal{O}_A) = \mathrm{Ext}(\mathcal{O}_A)$, $\mathrm{Ext}^0(\mathcal{O}_A) = \mathrm{Ext}(\mathcal{O}_A \otimes C_0(\mathbb{R}))$. The group $\mathrm{Ext}(\mathcal{O}_A)$ is an abelian group consisting of equivalence classes of short exact sequences of C^*-algebras:

$$0 \to \mathcal{K} \to E \to \mathcal{O}_A \to 0.$$

The crossed product $\mathcal{O}_A \rtimes_\alpha \mathbb{T}$ is stably isomorphic to the stabilization $\mathcal{F}_A \otimes \mathcal{K}$. The dual action $\hat{\alpha}$ of the gauge action corresponds to the shift σ_A on the AF-algebra, so that we have

COROLLARY 1.5. *The shift $\sigma_A = \hat{\alpha}_*$ on $K_0(\mathcal{F}_A)$ is an invariant for topological conjugacy.*

Cuntz and Krieger also proved that the isomorphism class of the C^*-algebra $\mathcal{O}_A \otimes \mathcal{K}$ is invariant under flow equivalence ([CK]). In particular, the group $\mathrm{Ext}(\mathcal{O}_A)$ is invariant under flow equivalence of the topological Markov shift (Λ_A, σ).

Let us describe a relationship between these K-theoretic objects of C^*-algebras and the preceding conjugacy invariants of topological Markovc shifts. In [CK], [C2], [C3], the following facts were proved:

THEOREM 1.6 (CUNTZ-KRIEGER [CK], CUNTZ [C2], [C3]).

$$(\Delta_A, \Delta_A^+, \delta_A) = (K_0(\mathcal{F}_A), K_0(\mathcal{F}_A)_+, \sigma_A),$$

$$BF(A) = \mathrm{Ext}(\mathcal{O}_A).$$

The dimension group of a matrix A is realized as the K_0-group of the AF-algebra $\mathcal{F}_A$. The Bowen-Franks group is realized as the Ext-group of the C^*-algebra $\mathcal{O}_A$ (cf. [H], [H2]). Krieger in [Kr2] and [Kr3] has defined the dimension group for A based on the idea of the computation: $K_0(\mathcal{F}_A) = \mathbb{Z}^n \xrightarrow{A^t} \mathbb{Z}^n \xrightarrow{A^t} \mathbb{Z}^n \xrightarrow{A^t} \cdots$. Incidentally, the term "dimension group "comes from the language of C^*-algebra ([Ef]). It is interesting that topological entropy of the topological Markov shift is realized as the inverse temperature of the admitted KMS-states for the gauge action on the Cuntz-Krieger algebra ([EFW3]).

For the classification problem of Cuntz-Krieger algebras, the isomorphism classes of Cuntz-Krieger algebras are completely classified by its K-groups. For the case that all entries of A are 1, that is, Λ_A is the full n-shift, the algebra $\mathcal{O}_A$ is the Cuntz algebra $\mathcal{O}_n$, whose K_0-group has been computed as $\mathbb{Z}^n/(n-1)\mathbb{Z}^n$ by Cuntz [C2]. It depends only on the size n of the matrix. For 3×3 irreducible matrices, Enomoto, Fujii, and Watatani have completely classified the algebras by means of K-groups, and also from a graph-theoretical viewpoint [EFW2]. For general irreducible matrices, M. Rørdam [Rø] completed their classsification. Recently, Bratteli and Kishimoto have proved that there exist one-to-one correspondences among the isomorphism class of the C^*-dynamical system $(\mathcal{O}_A \otimes \mathcal{K}, \alpha^A \otimes \mathrm{id}, \mathbb{T})$, the outer conjugacy class of the automorphism $\hat{\alpha}$ on the AF-algebra $\mathcal{F}_A \otimes \mathcal{K}$, and the shift equivalence class of the matrix ([BraKi]).

2. Subshifts and C^*-algebras

For general subshifts, the following problems are interesting from the symbolic dynamics viewpoint. How do we define dimension group and Bowen-Franks group as a topological conjugacy invariant and a flow equivalence invariant respectively? For what kind of objects do we formulate strong shift equivalence and shift equivalence? Is it possible to classify general subshifts by the strong shift equivalence classes? Concerning the structure theory of C^*-algebras, E. Kirchberg in the mid-90's has constructed a classification theory of unital, separable, nuclear, purely infinite, simple C^*-algebras [Kir], so we know that all such C^*-algebras are completely classified by their K-groups ([Kir], [Ph]). Hence, if we could generalize Cuntz-Krieger algebras for topological Markov shifts to C^*-algebras for subshifts, it would be possible to formulate topological conjugacy invariants and flow equivalence invariants for general subshifts by using the K-theoretic objects of these C^*-algebras. These C^*-algebras also would give good examples of Kirchberg's classifiable C^*-algebras.

With these motivations, we will try to construct C^*-algebras associated with subshifts. Let us recall the definition and notation of subshifts. Let Σ be an arbitrary finite set, which we call an alphabet. A member of Σ is called a symbol. For a closed shift-invariant (= σ-invariant) subset Λ of the infinite product space $\Sigma^{\mathbb{Z}}$, the topological dynamical system $(\Lambda, \sigma|_\Lambda)$ is called a subshift. A topological Markov shift is a subshift. We denote by X_Λ $(\subset \prod_{i=1}^{\infty} \Sigma_i)$ the set of all right one-sided sequences in Λ. The topological dynamical system $(X_\Lambda, \sigma|_{X_\Lambda})$ is called a right one-sided subshift. For a finite directed graph with finite vertices and finite edges, assign symbols of Σ to edges so that we have a labeled graph. We call such a graph a λ-graph. The set of all two-sided sequences of labels following the allowed directions of edges of the λ-graph yields a subshift, called a sofic shift, or originally a sofic system ([Fi], [We], [Kr4], [Kr5]). The class of sofic shifts properly contains the class of topological Markov shifts.

We henceforth fix an arbitrary subshift Λ. We may assume $\Sigma = \{1, 2, \dots, n\}$, $n > 1$. A finite sequence $\mu = (\mu_1, ..., \mu_k)$, $\mu_j \in \Sigma$, of Σ is called a word or a block. A word is said to be admissible if it appears in a place of a sequence of Λ. It is said to be forbidden otherwise. We may define a subshift by giving the set of all its admissible words, or the set of all its forbidden words. A subshift is said to be SFT if it has a finite set of forbidden words. The class of SFTs contains the class of topological Markov shifts and is contained in the class of sofic shifts.

We will now construct a C^*-algebra from a subshift in the following way (we will see in Section 4,5 that it is more natural to construct a C^*-algebra from a "presentation" of a subshift than from a subshift). We assign a vector e_x to each one-sided sequence $x \in X_\Lambda$. Let $\mathfrak{H}_\Lambda$ be the Hilbert space with its complete orthonomal basis $\{e_x | x \in X_\Lambda\}$. We define operators $S_1, \dots, S_n$ on $\mathfrak{H}_\Lambda$ by

$$S_j e_x = \begin{cases} e_{jx} & (jx \in X_\Lambda), \\ 0 & (jx \notin X_\Lambda). \end{cases}$$

We can immediately see that $S_1, \dots, S_n$ are partial isometries satisfying the relation $\sum_{j=1}^{n} S_j S_j^* = 1$.

Definition 2.1 (cf. [Ma]). The C^*-algebra $\mathcal{O}_\Lambda$ associated with subshift Λ is defined to be the C^*-algebra generated by the partial isometries $S_1, \dots, S_n$.

This definition is slightly different from the original definition given in [Ma]. In this article, we adopt this definition (cf. [CaM]). The algebraic structure of these C^*-algebras has been studied in [Ma], [Ma2], [Ma3]. If a subshift Λ is a topological Markov shift Λ_A, the C^*-algebra $\mathcal{O}_\Lambda$ is isomorphic to the Cuntz-Krieger algebra $\mathcal{O}_A$.

For a subshift Λ, let Λ^k be the set of all admissible words of length k. For a natural number $l \in \mathbb{N}$, we say that $x, y \in X_\Lambda$ are l-past equivalent if $\{\mu \in \Lambda^l | \mu x \in X_\Lambda\} = \{\nu \in \Lambda^l | \nu y \in X_\Lambda\}$. We denote this by $x \sim_l y$. We say that Λ satisfies condition (I) if for any $l \in \mathbb{N}$ and $x \in X_\Lambda$, there exists $y \in X_\Lambda$ such that $y \neq x$ and $y \sim_l x$. We henceforth assume condition (I) for subshifts. As well as a Cuntz-Krieger algebra, the correspondence $\alpha_t(S_i) = tS_i, i = 1, 2, \ldots, n, t \in \mathbb{T}$, gives rise to a one-parameter automorphism group on $\mathcal{O}_\Lambda$, called the gauge action.

THEOREM 2.2 ([MA], [MA3]). *If a subshift Λ is irreducible in past equivalence, the C^*-algebra $\mathcal{O}_\Lambda$ is simple. In particular, if Λ is aperiodic in past equivalence, $\mathcal{O}_\Lambda$ is simple and purely infinite.*

A subshift Λ is irreducible in past equivalence if for $l \in \mathbb{N}$, $y \in X_\Lambda$ and a sequence $x^k \in X_\Lambda$, $k \in \mathbb{N}$, satisfyng $x^k \sim_k x^{k+1}$, $k \in \mathbb{N}$, there exist a natural number N and an admissible word μ such that $y \sim_l \mu x^{l+N}$. A subshift Λ is aperiodic in past equivalence if for $l \in \mathbb{N}$ there exists a natural number N such that for any $x, y \in X_\Lambda$ there exists an admissible word μ of length N such that $y \sim_l \mu x^{l+N}$ ([Ma3]).

Thanks to this theorem, it is expected that various kinds of subshifts give rise to simple, purely infinite C^*-algebras. Recall that the Cuntz algebra $\mathcal{O}_n$ is associated to the full n-shift. For each real number $1 < \beta \in \mathbb{R}$, there is a subshift, called the β-shift, that comes from β-adic expansion of real numbers ([Pa], [Ré], [IT]). It is considered as a real number version of full n-shifts. Hence we may think of the C^*-algebras $\mathcal{O}_\beta, 1 < \beta \in \mathbb{R}$, associated with the β-shifts as interpolated Cuntz algebras. These C^*-algebras are simple and purely infinite by the above theorem ([KMW]).

Under a small condition on subshifts, we can show the following theorem, as a generalization of Theorem 1.3:

THEOREM 2.3 ([MA8], CF. [MA13]). *If two subshifts Λ_1 and Λ_2 are topologically conjugate, the stabilized C^*-algebras $\mathcal{O}_{\Lambda_1} \otimes \mathcal{K}$ and $\mathcal{O}_{\Lambda_2} \otimes \mathcal{K}$ with the gauge actions are isomorphic. That is, there exists an isomorphism ψ from $\mathcal{O}_{\Lambda_1} \otimes \mathcal{K}$ onto $\mathcal{O}_{\Lambda_2} \otimes \mathcal{K}$ such that $\psi \circ (\alpha_t^1 \otimes \mathrm{id}) = (\alpha_t^2 \otimes \mathrm{id}) \circ \psi, t \in \mathbb{T}$.*

Hence the C^*-dynamical system $(\mathcal{O}_\Lambda \otimes \mathcal{K}, \alpha \otimes \mathrm{id}, \mathbb{T})$ is an invariant of the subshift (Λ, σ) for topological conjugacy. The fixed point algebra ${\mathcal{O}_\Lambda}^\alpha$ of $\mathcal{O}_\Lambda$ under the gauge action α is an AF-algebra as well as a Cuntz-Krieger algebra. We denote it by $\mathcal{F}_\Lambda$. The crossed product $\mathcal{O}_\Lambda \rtimes_\alpha \mathbb{T}$ is stably isomorphic to $\mathcal{F}_\Lambda \otimes \mathcal{K}$. The above theorem implies that the dual action $\hat{\alpha}$ of α is invariant under topological conjugacy, so that one has

COROLLARY 2.4. *The action $\hat{\alpha}_*$ on the abelian group $K_0(\mathcal{F}_\Lambda)$ is invariant under topological conjugacy, and also the following abelian groups:*

$$K_0(\mathcal{O}_\Lambda), \quad K_1(\mathcal{O}_\Lambda), \quad \mathrm{Ext}^0(\mathcal{O}_\Lambda), \quad \mathrm{Ext}^1(\mathcal{O}_\Lambda)$$

are invariant under topological conjugacy.

If Λ is a topological Markov shift Λ_A, the group $\mathrm{Ext}(\mathcal{O}_A)$ coincides with the Bowen-Franks group $BF(A)$ of the matrix A, which is a flow equivalence invariant. We then define the Bowen-Franks group $BF(\Lambda)$ for a subshift Λ as

$$BF(\Lambda) = \mathrm{Ext}(\mathcal{O}_\Lambda)$$

(see [Ma10]).

THEOREM 2.5 ([MA10]). *The isomorphism class of the C^*-algebra $\mathcal{O}_\Lambda \otimes \mathcal{K}$ is invariant under flow equivalence. In particular, the group $BF(\Lambda)$ is a flow equivalence invariant.*

By extending a Parry-Sullivan argument [PS] for topological Markov shifts to subshifts, we can prove the above theorem by using Theorem 2.3 (cf. [Ma9]).

As well as the case of topological Markov shifts, the topological entropy for a subshift Λ appears as the inverse temperature of the admitted KMS-states for the gauge action on $\mathcal{O}_\Lambda$ under some condition ([MWY]). Pinzari, Watatani, and Yonetani have studied Voiculescu's noncommutative topological entropy and the Connes-Narnhofer-Thirring noncommutative measure-theoretical entropy for more general C^*-algebras constructed from Hilbert C^*-bimodules ([PWY], cf. [Ch], [BG], [KeP]). Automorphisms of the C^*-algebra $\mathcal{O}_\Lambda$ arising from automorphisms of subshifts have been studied in [Ma7] (cf. [EFW], [KT]).

3. Presentations of subshifts and strong shift equivalence

We have constructed some topological conjugacy invariants and flow equivalence invariants of subshifts by using C^*-algebras associated with the subshifts as in the previous section. Recalling the case of topological Markov shifts as in Section 1, we see that in fact the dimension group and the Bowen-Franks group have been defined for a matrix, not for a topological Markov shift. In particular, the dimension group plays an important rôle as a shift equivalence invariant of matrices. Moreover, topological Markov shifts and Cuntz-Krieger algebras are originally defined from matrices. A subshift is in general not given by a finite matrix.

We then introduce the notions of symbolic matrix system and λ-graph system, which are generalizations of the notions of symbolic matrix and λ-graph (= labeled graphs) respectively. For $l \in \mathbb{Z}_+ = \{0, 1, 2, \dots\}$, let $\mathcal{M}_{l,l+1}$ be a rectangular finite matrix with entries in formal sums of Σ and $I_{l,l+1}$ a rectangular finite matrix with entries in $\{0, 1\}$, whose size is the same as the size of $\mathcal{M}_{l,l+1}$. They satisfy the following relations:

$$I_{l,l+1}\mathcal{M}_{l+1,l+2} = \mathcal{M}_{l,l+1}I_{l+1,l+2}, \qquad l \in \mathbb{Z}_+.$$

Each row of $I_{l,l+1}$ has at least one 1, and each column has only one 1; also, each row and each column of $\mathcal{M}_{l,l+1}$ has at least one symbol. We call the sequences of the matrices $(\mathcal{M}_{l,l+1}, I_{l,l+1}), l \in \mathbb{Z}_+$ a symbolic matrix system ([Ma6]). It is the same as the following graphical object $\mathfrak{L} = (V, E, \lambda, \iota)$, called a λ-graph system. (V, E, λ) is a labeled Bratteli diagram with vertex set $V = V_0 \cup V_1 \cup V_2 \cup \cdots$ and directed edge set $E = E_{0,1} \cup E_{1,2} \cup E_{2,3} \cup \cdots$. An edge e of $E_{l,l+1}$ has its source $s(e)$ in V_l and its terminal $t(e)$ in V_{l+1}. It is labeled $\lambda(e) \in \Sigma$. ι consistis of a map $\iota_{l,l+1} : V_{l+1} \to V_l$ for each $l \in \mathbb{Z}_+$, corresponding to the matrix $I_{l,l+1}$. For $u \in V_l$ and $v \in V_{l+2}$ there is a bijective correspondence preserving the labels between the two sets

$$\{e \in E_{l,l+1} \mid s(e) = u, t(e) = \iota_{l+1,l+2}(v)\},$$
$$\{e \in E_{l+1,l+2} \mid \iota_{l,l+1}(s(e)) = u, t(e) = v\}.$$

We call this property the local property of the λ-graph system. We denote by $m(l)$ the number of the vertex set V_l. It is finite and satisfies $m(l) \le m(l+1)$ for each $l \in \mathbb{Z}_+$. These objects are generalizations of finite symbolic matrices and λ-graphs, respectively. For a square finite symbolic matrix $\mathcal{A}$, we set $\mathcal{M}_{l,l+1} = \mathcal{A}$ and $I_{l,l+1}$ the identity matrix with the same size as $\mathcal{A}$. The resulting pair becomes a symbolic matrix system. It is well-known that a finite symbolic matrix or a λ-graph defines a sofic shift. As a generalization of this fact, we may construct a subshift from a symbolic matrix system $(\mathcal{M}, I)$ in the following way. Let (V, E, λ, ι) be its λ-graph system. We put

$$X_{(\mathcal{M},I)} = \{(\lambda(e_1), \lambda(e_2), \dots) \in \Sigma^{\mathbb{N}} | e_i \in E_{i-1,i}, t(e_i) = s(e_{i+1}), \forall i \in \mathbb{N}\},$$

the set of all right one-sided sequences of labels appearing in the labeled Bratteli diagram. It is a compact Hausdorff space in the product topology. By the local property of the λ-graph system, for $(\alpha_1, \alpha_2, \dots) \in X_{(\mathcal{M},I)}$ one sees that $(\alpha_2, \alpha_3, \dots) \in X_{(\mathcal{M},I)}$ and there exists $\alpha_0 \in \Sigma$ such that $(\alpha_0, \alpha_1, \alpha_2, \dots) \in X_{(\mathcal{M},I)}$. The shift

$$S : (\alpha_1, \alpha_2, \alpha_3, \dots) \in X_{(\mathcal{M},I)} \to (\alpha_2, \alpha_3, \dots) \in X_{(\mathcal{M},I)}$$

is a surjective and continuous map. We set

$$\Lambda_{(\mathcal{M},I)} = \varprojlim \{S : X_{(\mathcal{M},I)} \to X_{(\mathcal{M},I)}\},$$

the projective limit by the shift; that is,

$$\Lambda_{(\mathcal{M},I)} = \{(\dots, \alpha_{-2}, \alpha_{-1}, \alpha_0, \alpha_1, \alpha_2, \dots) | (\alpha_n, \alpha_{n+1}, \dots) \in X_{(\mathcal{M},I)}, \forall n \in \mathbb{Z}\}.$$

It then follows that S defines a homeomorphism $\sigma((\alpha_i)_{i\in\mathbb{Z}}) = (\alpha_{i+1})_{i\in\mathbb{Z}}$ so that we obtain a subshift $(\Lambda_{(\mathcal{M},I)}, \sigma)$. Conversely, an arbitrary subshift Λ gives rise to a symbolic matrix system in the following way. Let $F_i^l, i = 1, 2, \dots, m(l)$, be the l-past equivalence classes of X_Λ. For $i = 1, 2, \dots, m(l)$ and $j = 1, 2, \dots, m(l+1)$ the (i,j)-component $I_{l,l+1}^\Lambda(i,j)$ of $I_{l,l+1}^\Lambda$ is defined to be 1 if $F_j^{l+1} \subset F_i^l$, otherwise 0. The (i,j)-component $\mathcal{M}_{l,l+1}^\Lambda(i,j)$ of $\mathcal{M}_{l,l+1}^\Lambda$ is defined to be $a_1 + \dots + a_m$, where $\{a_1, \dots, a_m\} = \{a \in \Sigma \mid \exists x \in F_j^{l+1}; ax \in F_i^l\}$. One then sees that the pair $(\mathcal{M}^\Lambda, I^\Lambda)$ becomes a symbolic matrix system. We call it the canonical symbolic matrix system for Λ. Its corresponding λ-graph system is called the canonical λ-graph system for Λ. If in particular, a subshift is a sofic shift, the canonical λ-graph system corresponds to its left Krieger cover graph.

M. Nasu in [N], [N2] introduced the notion of strong shift equivalence for symbolic matrices, and generalized Theorem 1.1 to sofic shifts. We will introduce the notion of strong shift equivalence for symbolic matrix systems, and generalize Nasu's result to subshifts. For two symbolic matrices $\mathcal{A}, \mathcal{A}'$ over Σ, Σ' respectively, we say that $\mathcal{A}$ and $\mathcal{A}'$ are equivalent under specification ϕ if there exists a bijection ϕ from a subset of Σ to a subset of Σ' such that the matrix $\mathcal{A}'$ is obtained by replacing all the symbols a in $\mathcal{A}$ by $\phi(a)$. We write this as $\mathcal{A} \overset{\phi}{\simeq} \mathcal{A}'$ ([N], [N2]). For alphabets C, D, we denote by $C \cdot D$ the set of all finite formal sums of formal products of elements of C and D. Let $(\mathcal{M}, I)$ and $(\mathcal{M}', I')$ be symbolic matrix systems over Σ

and Σ' respectively, where both $\mathcal{M}_{l,l+1}$ and $I_{l,l+1}$ are $m(l) \times m(l+1)$ matrices and both $\mathcal{M}'_{l,l+1}$ and $I'_{l,l+1}$ are $m'(l) \times m'(l+1)$ matrices.

DEFINITION 3.1. If there exist alphabets C, D, specifications $\varphi : \Sigma \to C \cdot D, \phi : \Sigma' \to D \cdot C$, and, for $l \in \mathbb{N}$, an $m(l-1) \times m'(l)$ symbolic matrix $\mathcal{H}_l$ over C and an $m'(l-1) \times m(l)$ symbolic matrix $\mathcal{K}_l$ over D such that

$$I_{l-1,l}\mathcal{M}_{l,l+1} \overset{\varphi}{\simeq} \mathcal{H}_l\mathcal{K}_{l+1}, \qquad I'_{l-1,l}\mathcal{M}'_{l,l+1} \overset{\phi}{\simeq} \mathcal{K}_l\mathcal{H}_{l+1},$$

$$\mathcal{H}_l I'_{l,l+1} = I_{l-1,l}\mathcal{H}_{l+1}, \qquad \mathcal{K}_l I_{l,l+1} = I'_{l-1,l}\mathcal{K}_{l+1},$$

then we say that $(\mathcal{M}, I)$ and $(\mathcal{M}', I')$ are strong shift equivalent in 1 step, and write it as $(\mathcal{M}, I) \underset{\text{1-st}}{\approx} (\mathcal{M}', I')$. If there exists a finite sequence of symbolic matrix systems $(\mathcal{M}^{(i)}, I^{(i)}), i = 1, 2, \dots, N-1$, such that

$$(\mathcal{M}, I) \underset{\text{1-st}}{\approx} (\mathcal{M}^{(1)}, I^{(1)}) \underset{\text{1-st}}{\approx} (\mathcal{M}^{(2)}, I^{(2)}) \underset{\text{1-st}}{\approx} \cdots \underset{\text{1-st}}{\approx} (\mathcal{M}^{(N-1)}, I^{(N-1)}) \underset{\text{1-st}}{\approx} (\mathcal{M}', I'),$$

then we say that $(\mathcal{M}, I)$ and $(\mathcal{M}', I')$ are strong shift equivalent, and write it as $(\mathcal{M}, I) \approx (\mathcal{M}', I')$.

It can be proved that

THEOREM 3.2 ([MA6]). *Two subshifts Λ and Λ' are topologically conjugate if and only if their canonical symbolic matrix systems $(\mathcal{M}^{\Lambda}, I^{\Lambda})$ and $(\mathcal{M}^{\Lambda'}, I^{\Lambda'})$ are strong shift equivalent.*

It is immediate to see that if $(\mathcal{M}^{\Lambda}, I^{\Lambda})$ and $(\mathcal{M}^{\Lambda'}, I^{\Lambda'})$ are strong shift equivalent, the subshifts Λ and Λ' are topologically conjugate. For the proof of the converse implication, we need the notion of bipartite λ-graph system so that we may prove it by using Nasu's factorization theorem of topological conjugacy into bipartite conjugacies (cf. [N]). We may also directly prove the converse implication by considering state splits and state amalgamations of λ-graph systems ([Ma15]).

We will introduce the notion of shift equivalence of symbolic matrix systems. For $N \in \mathbb{N}$, we denote by $(\Sigma)^N = \Sigma \cdots \Sigma, (\Sigma')^N = \Sigma' \cdots \Sigma'$ (N-times products).

DEFINITION 3.3. If there exist alphabets C_N, D_N, specifications $\varphi_1 : \Sigma \cdot C_N \to C_N \cdot \Sigma', \varphi_2 : \Sigma' \cdot D_N \to D_N \cdot \Sigma$ and $\psi_1 : (\Sigma)^N \to C_N \cdot D_N, \psi_2 : (\Sigma')^N \to D_N \cdot C_N$, and for $l \in \mathbb{Z}_+$, an $m(l) \times m'(l+N)$ symbolic matrix $\mathcal{H}_l$ over C_N and an $m'(l) \times m(l+N)$ symbolic matrix $\mathcal{K}_l$ over D_N such that

$$\mathcal{M}_{l,l+1}\mathcal{H}_{l+1} \overset{\varphi_1}{\simeq} \mathcal{H}_l\mathcal{M}'_{l+N,l+N+1}, \qquad \mathcal{M}'_{l,l+1}\mathcal{K}_{l+1} \overset{\varphi_2}{\simeq} \mathcal{K}_l\mathcal{M}_{l+N,l+N+1},$$

$$I_{l,l+N}\mathcal{M}_{l+N,l+2N} \overset{\psi_1}{\simeq} \mathcal{H}_l\mathcal{K}_{l+N}, \qquad I'_{l,l+N}\mathcal{M}'_{l+N,l+2N} \overset{\psi_2}{\simeq} \mathcal{K}_l\mathcal{H}_{l+N},$$

$$I_{l,l+1}\mathcal{H}_{l+1} = \mathcal{H}_l I'_{l+N,l+N+1}, \qquad I'_{l,l+1}\mathcal{K}_{l+1} = \mathcal{K}_l I_{l+N,l+N+1},$$

then the symbolic matrix systems $(\mathcal{M}, I)$ and $(\mathcal{M}', I')$ are said to be shift equivalent, and we write it as $(\mathcal{M}, I) \sim (\mathcal{M}', I')$.

Strong shift equivalence implies shift equivalence. Shift equivalence for symbolic matrices was formulated by Boyle and Krieger [BK]. The above formulation is a generalization of the Boyle-Krieger formulation for symbolic matrices.

As seen above, we have extended the notion of strong shift equivalence and shift equivalence to symbolic matrix systems, related to topological conjugacy of subshifts. Since the components of the symbolic matrices appearing in symbolic matrix systems consist of symbols, we cannot formulate the dimension groups and the Bowen-Franks groups as they are. We then introduce the notion of nonnegative matrix system to yield computable objects. If rectangular nonnegative matrices $A_{l,l+1}$ and $\{0,1\}$-matrices $I_{l,l+1}$ for $l \in \mathbb{Z}_+$ satisfy the relations

$$I_{l,l+1}A_{l+1,l+2} = A_{l,l+1}I_{l+1,l+2}, \qquad l \in \mathbb{Z}_+, \tag{3.1}$$

then the sequence $(A_{l,l+1}, I_{l,l+1}), l \in \mathbb{Z}_+$, of the pairs is called a nonnegative matrix system ([Ma6]). For a symbolic matrix system $(\mathcal{M}, I)$, let $M_{l,l+1}$ be the nonnegative matrix obtained by setting all the symbols of the components of $\mathcal{M}_{l,l+1}$ equal to 1. Then $(M_{l,l+1}, I_{l,l+1}), l \in \mathbb{Z}_+$, becomes a nonnegative matrix system. We may formulate strong shift equivalence and shift equivalence for nonnegative matrix systems by replacing symbolic matrices and specifications by nonnegative matrices and equals in Definitions 3.1 and 3.3. We formulate the dimension groups and the Bowen-Franks groups for nonnegative matrix systems in the next section.

4. Dimension groups and Bowen-Franks groups for nonnegative matrix systems

Let (A, I) be a nonnegative matrix system. Define the inductive limits:

$$\mathbb{Z}_{I^t} = \varinjlim_{l}\{I^t_{l,l+1} : \mathbb{Z}^{m(l)} \to \mathbb{Z}^{m(l+1)}\},$$
$$\mathbb{Z}^+_{I^t} = \varinjlim_{l}\{I^t_{l,l+1} : \mathbb{Z}^{m(l)}_+ \to \mathbb{Z}^{m(l+1)}_+\},$$

where $I^t_{l,l+1}$ is the transpose of $I_{l,l+1}$. By (3.1), the sequence of the transposes $A^t_{l,l+1}, l \in \mathbb{Z}_+$, of $A_{l,l+1}$ naturally acts on the ordered group $(\mathbb{Z}_{I^t}, \mathbb{Z}^+_{I^t})$ and defines an endomorphism on it. We denote it by $\lambda_{(A,I)}$. We put $\mathbb{Z}_{I^t}(k) = \mathbb{Z}_{I^t}, k \in \mathbb{N}$, and consider the inductive limits

$$\Delta_{(A,I)} = \varinjlim_{k}\{\lambda_{(A,I)} : \mathbb{Z}_{I^t}(k) \to \mathbb{Z}_{I^t}(k+1)\},$$
$$\Delta^+_{(A,I)} = \varinjlim_{k}\{\lambda_{(A,I)} : \mathbb{Z}^+_{I^t}(k) \to \mathbb{Z}^+_{I^t}(k+1)\}.$$

The ordered group $(\Delta_{(A,I)}, \Delta^+_{(A,I)})$ is called the dimension group for (A, I). The map $\delta_{(A,I)} : \mathbb{Z}_{I^t}(k) \to \mathbb{Z}_{I^t}(k+1)$ defined by $\delta_{(A,I)}([X,k]) = ([X,k+1])$ yields an automorphism on $(\Delta_{(A,I)}, \Delta^+_{(A,I)})$. The triple $(\Delta_{(A,I)}, \Delta^+_{(A,I)}, \delta_{(A,I)})$ is called the dimension triple for (A, I).

PROPOSITION 4.1. *The dimension triple $(\Delta_{(A,I)}, \Delta^+_{(A,I)}, \delta_{(A,I)})$ for (A, I) is an invariant for shift equivalence.*

For $l \in \mathbb{Z}_+$, we put

$$K^l_0(A, I) = \mathbb{Z}^{m(l+1)}/(I^t_{l,l+1} - A^t_{l,l+1})\mathbb{Z}^{m(l)},$$
$$K^l_1(A, I) = \mathrm{Ker}(I^t_{l,l+1} - A^t_{l,l+1}).$$

The map $I^t_{l,l+1} : \mathbb{Z}^{m(l)} \to \mathbb{Z}^{m(l+1)}$ naturally extends to homomorphisms

$$i^l_* : K^l_*(A, I) \to K^{l+1}_*(A, I), \qquad * = 0, 1.$$

We consider their inductive limits

$$K_*(A,I) = \varinjlim_l \{i_*^l : K_*^l(A,I) \to K_*^{l+1}(A,I)\}, \qquad * = 0,1,$$

and call them the K-groups for (A,I). They are realized as the cokernel and the kernel of the endomorphism id $-\ \delta_{(A,I)}$ on the group $\Delta_{(A,I)}$, respectively. The abelian group $\mathbb{Z}_I$ is defined by the projective limit:

$$\mathbb{Z}_I = \varprojlim_l \{I_{l,l+1} : \mathbb{Z}^{m(l+1)} \to \mathbb{Z}^{m(l)}\}.$$

The sequence of the matrices $A_{l,l+1}, l \in \mathbb{Z}_+$, gives rise to an endomorphism on $\mathbb{Z}_I$. We denote it by A_I. The Bowen-Franks groups for (A,I) are defined by

$$BF^0(A,I) = \mathbb{Z}_I/(I - A_I)\mathbb{Z}_I, \qquad BF^1(A,I) = \mathrm{Ker}(I - A_I).$$

By an elementary calculation of homological algebra, we have a short exact sequence

$$0 \longrightarrow \mathrm{Ext}^1_{\mathbb{Z}}(K_0(A,I),\mathbb{Z}) \xrightarrow{\delta} BF^0(A,I) \xrightarrow{\gamma} \mathrm{Hom}_{\mathbb{Z}}(K_1(A,I),\mathbb{Z}) \longrightarrow 0$$

that splits unnaturally. We also have an isomorphism

$$BF^1(A,I) \cong \mathrm{Hom}_{\mathbb{Z}}(K_0(A,I),\mathbb{Z}).$$

Although the above short exact sequence is a purely algebraic setting, it corresponds to the universal coefficient theorem for K-theory of C^*-algebras ([Bro], [RS]). We summarise the above discussions and reach the following theorem.

THEOREM 4.2 ([MA6]). *Let $(\mathcal{M},I), (\mathcal{M}',I')$ be symbolic matrix systems, and let (M,I), (M',I') be their respective nonnegative matrix systems. Consider the following situations:*

(S1) $(\mathcal{M},I) \approx (\mathcal{M}',I')$ *: strong shift equivalence,*
(N1) $(M,I) \approx (M',I')$ *: strong shift equivalence,*
(S2) $(\mathcal{M},I) \sim (\mathcal{M}',I')$ *: shift equivalence,*
(N2) $(M,I) \sim (M',I')$ *: shift equivalence,*
(3) $(\Delta_{(M,I)}, \Delta^+_{(M,I)}, \delta_{(M,I)}) \cong (\Delta_{(M',I')}, \Delta^+_{(M',I')}, \delta_{(M',I')})$ *: isomorphic dimension triples,*
(4) $(\Delta_{(M,I)}, \delta_{(M,I)}) \cong (\Delta_{(M',I')}, \delta_{(M',I')})$ *: isomorphic dimension groups as groups with dimension automorphisms,*
(5) $K_*(M,I) \cong K_*(M',I')$ *: isomorphic K-groups,*
(6) $BF^*(M,I) \cong BF^*(M',I')$ *: isomorphic Bowen-Franks groups.*

Then the following implications hold:

$$\begin{array}{ccccccccccc} (S1) & \Longrightarrow & (S2) & & & & & & & & \\ \Downarrow & & \Downarrow & & & & & & & & \\ (N1) & \Longrightarrow & (N2) & \Longrightarrow & (3) & \Longrightarrow & (4) & \Longrightarrow & (5) & \Longrightarrow & (6). \end{array}$$

Relationships to topological conjugacy invariants and flow equivalence invariants of subshifts defined by using C^*-algebras in Section 3 are the following:

THEOREM 4.3 ([MA2], [MA3], [MA6], [MA10], CF. [C3], [CK]). *For a subshift Λ, let (M,I) be the nonnegative matrix system defined from the canonical symbolic matrix system $(\mathcal{M},I)$ for Λ. The following isomorphisms hold:*

$$
\begin{aligned}
(\Delta_{(M,I)}, \Delta^+_{(M,I)}, \delta_{(M,I)}) &\cong (K_0(\mathcal{F}_\Lambda), K_0(\mathcal{F}_\Lambda)_+, \hat{\alpha_{\Lambda}}_*),\\
K_i(M,I) &\cong K_i(\mathcal{O}_\Lambda), \qquad i=0,1,\\
BF^i(M,I) &\cong \mathrm{Ext}^{i+1}(\mathcal{O}_\Lambda), \qquad i=0,1.
\end{aligned}
$$

We may give an isomorphism from $BF^0(M,I)$ onto $\mathrm{Ext}(\mathcal{O}_\Lambda)$ by computing the analytic indices of Fredholm operators associated with a representation of the C^*-algebra $\mathcal{O}_\Lambda$ on a Hilbert space ([Ma10]). This discussion is due to a generalized argument from Cuntz and Krieger's proof ([CK]). In [Ma9], it is proved that the group $BF(M,I)$ is invariant under flow equivalence of subshifts by a generalized argument following Bowen and Franks's proof in [BF]. The proof does not need C^*-algebras.

The eigenvectors and the eigenvalues for a nonnegative matrix system have been introduced in [Ma6]. They exactly correspond to the KMS-states and the inverse temperatures for the gauge action on the C^*-algebra, respectively ([MWY]).

5. The C^*-algebras associated with presentations of subshifts

The symbolic matrix system $(\mathcal{M},I)$ in Theorem 4.3 comes from the canonical λ-graph system for a subshift. In this section, we will construct C^*-algebras from λ-graph systems that are not necessarily canonical, as a generalization of the C^*-algebras associated with subshifts. We will first construct a continuous graph from a λ-graph system in the sense of V. Deaconu ([De2], [De3], [De4]). Let $\mathfrak{L}=(V,E,\lambda,\iota)$ be a λ-graph system over Σ and $(\mathcal{M},I)$ its symbolic matrix system. We denote by $(M,I)=(M_{l,l+1},I_{l,l+1})_{l\in\mathbb{Z}_+}$ its nonnegative matrix system. We assume that $\mathfrak{L}$ is left-resolving: for $e,e'\in E_{l,l+1}$, the equalities $t(e)=t(e')$ and $\lambda(e)=\lambda(e')$ imply $e=e'$. We consider the projective limit

$$\Omega_{\mathfrak{L}}=\varprojlim V_l=\{(v^l)_{l\in\mathbb{Z}_+}\in\prod_{l\in\mathbb{Z}_+}V_l \mid \iota_{l,l+1}(v^{l+1})=v^l, l\in\mathbb{Z}_+\}$$

of the system $\iota_{l,l+1}:V_{l+1}\to V_l, l\in\mathbb{Z}_+$. It becomes a compact Hausdorff space in a natural way. A member of $\Omega_{\mathfrak{L}}$ is called an ι-orbit. Set

$$
\begin{aligned}
E_{\mathfrak{L}}=\{(u,\alpha,v)\in\Omega_{\mathfrak{L}}\times\Sigma\times\Omega_{\mathfrak{L}} \mid u=(u^l)_{l\in\mathbb{Z}_+}, v=(v^l)_{l\in\mathbb{Z}_+},\\
\forall l\in\mathbb{Z}_+, \exists e_{l,l+1}\in E_{l,l+1}; u^l=s(e_{l,l+1}), v^{l+1}=t(e_{l,l+1}), \alpha=\lambda(e_{l,l+1})\}.
\end{aligned}
$$

As $E_{\mathfrak{L}}$ is a closed subset of $\Omega_{\mathfrak{L}}\times\Sigma\times\Omega_{\mathfrak{L}}$, it is a continuous graph in the sense of Deaconu ([De2], [De3], [De4]). Following his construction, consider the set of all right one-sided infinite paths of $E_{\mathfrak{L}}$:

$$
\begin{aligned}
X_{\mathfrak{L}}=\{(\alpha_i,v_i)_{i=1}^\infty\in\prod_{i=1}^\infty(\Sigma\times\Omega_{\mathfrak{L}}) \mid \exists v_0\in\Omega_{\mathfrak{L}}; (v_0,\alpha_1,v_1)\in E_{\mathfrak{L}},\\
(v_i,\alpha_{i+1},v_{i+1})\in E_{\mathfrak{L}}, \forall i\in\mathbb{N}\}.
\end{aligned}
$$

The shift map $\sigma:(\alpha_i,u_i)_{i=1}^\infty\in X_{\mathfrak{L}}\to(\alpha_{i+1},u_{i+1})_{i=1}^\infty\in X_{\mathfrak{L}}$ is a local homeomorphism on $X_{\mathfrak{L}}$. One may construct a locally compact r-discrete groupoid from the local homeomorphism σ on $X_{\mathfrak{L}}$ by the discussions of [De2], [De3], [De3] (cf. [An], [Re]). Set

$$G_{\mathfrak{L}} = \{(x,n,y) \in X_{\mathfrak{L}} \times \mathbb{Z} \times X_{\mathfrak{L}} \mid \exists k,l \geq 0;\ \sigma^k(x) = \sigma^l(y), n = k-l\}.$$

The range map and the domain map are defined by

$$r(x,n,y) = x, \qquad d(x,n,y) = y.$$

The multiplication and the inverse operation are defined by

$$(x,n,y)(y,m,z) = (x,n+m,z), \qquad (x,n,y)^{-1} = (y,-n,x).$$

The unit space $G^0_{\mathfrak{L}}$ is defined to be the space $X_{\mathfrak{L}} = \{(x,0,x) \in G_{\mathfrak{L}} \mid x \in X_{\mathfrak{L}}\}$. A basis of the open sets for $G_{\mathfrak{L}}$ is given by

$$Z(U,V,k,l) = \{(x,k-l,(\sigma^l|_V)^{-1} \circ (\sigma^k(x)) \in G_{\mathfrak{L}} \mid x \in U\},$$

where U,V are open sets of $X_{\mathfrak{L}}$, and $k,l \in \mathbb{N}$ are such that $\sigma^k|_U$ and $\sigma^l|_V$ are homeomorphisms with the same open range. Hence we see that

$$Z(U,V,k,l) = \{(x,k-l,y) \in G_{\mathfrak{L}} \mid x \in U, y \in V, \sigma^k(x) = \sigma^l(y)\}.$$

The groupoid C^*-algebra $C^*(G_{\mathfrak{L}})$ for the groupoid $G_{\mathfrak{L}}$ is defined as in the following way ([Re], cf. [An], [De2], [De3], [De4]). Let $C_c(G_{\mathfrak{L}})$ be the set of all continuous functions on $G_{\mathfrak{L}}$ with compact support; it has a natural product structure of $*$-algebra given by

$$(f*g)(s) = \sum_{\substack{t\in G_{\mathfrak{L}},\\ r(t)=r(s)}} f(t)g(t^{-1}s) = \sum_{\substack{t_1,t_2\in G_{\mathfrak{L}},\\ s=t_1t_2}} f(t_1)g(t_2),$$
$$f^*(s) = \overline{f(s^{-1})}, \qquad f,g \in C_c(G_{\mathfrak{L}}), \quad s \in G_{\mathfrak{L}}.$$

Let $C_0(G^0_{\mathfrak{L}})$ be the C^*-algebra of all continuous functions on $G^0_{\mathfrak{L}}$ that vanish at infinity. The algebra $C_c(G_{\mathfrak{L}})$ is a $C_0(G^0_{\mathfrak{L}})$-module, endowed with a $C_0(G^0_{\mathfrak{L}})$-valued inner product by

$$(\xi f)(x,n,y) = \xi(x,n,y)f(y), \qquad \xi \in C_c(G_{\mathfrak{L}}),\quad f \in C_0(G^0_{\mathfrak{L}}),\quad (x,n,y) \in G_{\mathfrak{L}},$$
$$\langle \xi,\eta\rangle(y) = \sum_{\substack{x,n\\ (x,n,y)\in G_{\mathfrak{L}}}} \overline{\xi(x,n,y)}\eta(x,n,y), \qquad \xi,\eta \in C_c(G_{\mathfrak{L}}),\quad y \in G^0_{\mathfrak{L}} (= X_{\mathfrak{L}}).$$

Let us denote by $l^2(G_{\mathfrak{L}})$ the completion of the inner product $C_0(G^0_{\mathfrak{L}})$-module $C_c(G_{\mathfrak{L}})$. It is a Hilbert C^*-right module over the commutative C^*-algebra $C_0(G^0_{\mathfrak{L}})$. We denote by $B(l^2(G_{\mathfrak{L}}))$ the C^*-algebra of all bounded adjointable $C_0(G^0_{\mathfrak{L}})$-module maps on $l^2(G_{\mathfrak{L}})$. Let π be the $*$-homomorphism of $C_c(G_{\mathfrak{L}})$ into $B(l^2(G_{\mathfrak{L}}))$ defined by $\pi(f)\xi = f * \xi$ for $f,\xi \in C_c(G_{\mathfrak{L}})$. Then the closure of $\pi(C_c(G_{\mathfrak{L}}))$ in $B(l^2(G_{\mathfrak{L}}))$ is called the (reduced) C^*-algebra of the groupoid $G_{\mathfrak{L}}$, which we denote by $C^*(G_{\mathfrak{L}})$.

DEFINITION 5.1 ([MA12]). The C^*-algebra $\mathcal{O}_{\mathfrak{L}}$ associated with the λ-graph system $\mathfrak{L}$ is defined to be the C^*-algebra $C^*(G_{\mathfrak{L}})$ of the groupoid $G_{\mathfrak{L}}$.[1]

[1]The construction of the C^*-algebra $\mathcal{O}_{\mathfrak{L}}$ is slightly different from the construction given in the original paper. By Theorem 5.2, the resulting C^*-algebras for predecessor-separated λ-graph systems are the same as the C^*-algebras constructed in the original paper.

Let $\{\mathfrak{v}_1^l, \dots, \mathfrak{v}_{m(l)}^l\}$ be the vertex set V_l. For a vertex $\mathfrak{v}_i^l \in V_l$, let $\Gamma^+(\mathfrak{v}_i^l)$ be the set of all sequences of labels in $\mathfrak{L}$ starting at $\mathfrak{v}_i^l$. That is,

$$\begin{aligned}\Gamma^+(\mathfrak{v}_i^l) = \{(\alpha_1, \alpha_2, \dots,) \in \Sigma^{\mathbb{N}} \mid \exists e_{n,n+1} \in E_{n,n+1} \text{ for } n = l, l+1, \dots; \\ \mathfrak{v}_i^l = s(e_{l,l+1}),\ t(e_{n,n+1}) = s(e_{n+1,n+2}),\ \lambda(e_{n,n+1}) = \alpha_{n-l+1}\}.\end{aligned}$$

We say that $\mathfrak{L}$ satisfies condition (I) if for each $\mathfrak{v}_i^l \in V$, the set $\Gamma^+(\mathfrak{v}_i^l)$ contains at least two distinct sequences. In this case $\mathcal{O}_{\mathfrak{L}}$ is the unique C^*-algebra subject to operator relations encoded in the structure of the λ-graph system among generating partial isometries corresponding to the labels and projections corresponding to the vertices in the following way.

THEOREM 5.2 ([MA12]). *Suppose that a λ-graph system $\mathfrak{L}$ satisfies condition (I). Then the C^*-algebra $\mathcal{O}_{\mathfrak{L}}$ is the unique universal concrete C^*-algebra generated by partial isometries $S_\alpha, \alpha \in \Sigma$, and projections $E_i^l, i = 1, 2, \dots, m(l)$, $l \in \mathbb{Z}_+$, subject to the following operator relations:*

$$\sum_{\alpha\in\Sigma} S_\alpha S_\alpha^* = 1, \tag{5.1}$$

$$\sum_{i=1}^{m(l)} E_i^l = 1, \qquad E_i^l = \sum_{j=1}^{m(l+1)} I_{l,l+1}(i,j) E_j^{l+1}, \tag{5.2}$$

$$S_\alpha S_\alpha^* E_i^l = E_i^l S_\alpha S_\alpha^*, \tag{5.3}$$

$$S_\alpha^* E_i^l S_\alpha = \sum_{j=1}^{m(l+1)} A_{l,l+1}(i,\alpha,j) E_j^{l+1}, \tag{5.4}$$

for $i = 1, 2, \dots, m(l), l \in \mathbb{Z}_+, \alpha \in \Sigma$, where

$$A_{l,l+1}(i,\alpha,j) = \begin{cases} 1 & \text{if } s(e) = \mathfrak{v}_i^l, \lambda(e) = \alpha, t(e) = \mathfrak{v}_j^{l+1} \text{ for some } e \in E_{l,l+1}, \\ 0 & \text{otherwise,} \end{cases}$$

$$I_{l,l+1}(i,j) = \begin{cases} 1 & \text{if } \iota_{l,l+1}(\mathfrak{v}_j^{l+1}) = \mathfrak{v}_i^l, \\ 0 & \text{otherwise} \end{cases}$$

for $i = 1, 2, \dots, m(l),\ j = 1, 2, \dots, m(l+1),\ \alpha \in \Sigma$.

For $x = (\alpha_i, u_i)_{i=1}^\infty \in X_{\mathfrak{L}}$, we put $\lambda(x)_1 = \alpha_1 \in \Sigma$. Now $\mathfrak{L}$ is left-resolving, so there exists a unique $v(x)_0 = (v(x)_0^l)_{l\in\mathbb{Z}_+} \in \Omega_{\mathfrak{L}} = \varprojlim V_l$ satisfying $(v(x)_0, \alpha_1, u_1) \in E_{\mathfrak{L}}$. Set, for $\alpha \in \Sigma$, $\mathfrak{v}_i^l \in V_l$,

$$\begin{aligned} U(\alpha) =& \{(x,1,y) \in G_{\mathfrak{L}} \mid \sigma(x) = y, \lambda(x)_1 = \alpha\}, \\ U(\mathfrak{v}_i^l) =& \{(x,0,x) \in G_{\mathfrak{L}} \mid v(x)_0^l = \mathfrak{v}_i^l\}. \end{aligned}$$

They are clopen sets of $G_{\mathfrak{L}}$. By setting

$$S_\alpha = \pi(\chi_{U(\alpha)}), \qquad E_i^l = \pi(\chi_{U(\mathfrak{v}_i^l)}) \qquad \text{in} \qquad \pi(C_c(G_{\mathfrak{L}})),$$

where $\chi_F \in C_c(G_{\mathfrak{L}})$ denotes the characteristic function of a clopen set F on the space $G_{\mathfrak{L}}$, one sees that they generate $\mathcal{O}_{\mathfrak{L}}$ and satisfy (5.1)–(5.4).

We remark that (5.3) is equivalent to the relation

$$E_i^l = \sum_{\alpha\in\Sigma}\sum_{j=1}^{m(l+1)} A_{l,l+1}(i,\alpha,j)S_\alpha E_j^{l+1}S_\alpha^* \tag{5.5}$$

under the relations (5.1) and (5.4).[2] We denote by $\Gamma_l^-(\mathfrak{v}_i^l)$ the set of all words presented by paths starting at a vertex of V_0 and terminating at the vertex $\mathfrak{v}_i^l$. $\mathfrak{L}$ is said to be predecessor-separated if $\Gamma_l^-(\mathfrak{v}_i^l) \neq \Gamma_l^-(\mathfrak{v}_j^l)$ for distinct i,j and for $l \in \mathbb{N}$. In this case, the relations

$$E_i^l = \prod_{\substack{\mu,\nu\in\Sigma^l \\ \mu\in\Gamma_l^-(\mathfrak{v}_i^l),\nu\notin\Gamma_l^-(\mathfrak{v}_i^l)}} S_\mu^* S_\mu(1 - S_\nu^* S_\nu), \qquad l \in \mathbb{N},$$

$$E_i^0 = \sum_{\alpha\in\Sigma}\sum_{j=1}^{m(1)} A_{0,1}(i,\alpha,j)S_\alpha E_j^1 S_\alpha^*$$

hold for $i = 1,2,\dots,m(l)$, where $S_\mu = S_{\mu_1}\cdots S_{\mu_k}$ for $\mu = (\mu_1,\dots,\mu_k) \in \Sigma^k$. Hence, if $\mathfrak{L}$ is predecessor-separated, $\mathcal{O}_{\mathfrak{L}}$ is generated by only the partial isometries $S_\alpha, \alpha \in \Sigma$.

For $x = (\alpha_i, v_i)_{i=1}^\infty \in X_{\mathfrak{L}}$, put $v_0 = v(x)_0 \in \Omega_{\mathfrak{L}}$. For a symbol $\beta \in \Sigma$, if there exists $v_{-1} \in \Omega_{\mathfrak{L}}$ satisfying $(v_{-1}, \beta, v_0) \in E_{\mathfrak{L}}$, we define $\beta x \in X_{\mathfrak{L}}$, by putting $\alpha_0 = \beta$, as

$$\beta x = (\alpha_{i-1}, v_{i-1})_{i=1}^\infty \in X_{\mathfrak{L}}.$$

Similarly to the construction of the C^*-algebras associated with subshifts as in Section 2, let e_x be a vector assigned to $x \in X_{\mathfrak{L}}$. Let $\mathfrak{H}_{\mathfrak{L}}$ be the Hilbert space spanned by the vectors $e_x, x \in X_{\mathfrak{L}}$, such that the vectors $e_x, x \in X_{\mathfrak{L}}$, form its complete orthonormal basis. Put

$$\Gamma_1^-(x) = \{\gamma \in \Sigma \mid (v_{-1}, \gamma, v(x)_0) \in E_{\mathfrak{L}} \text{ for some } v_{-1} \in \Omega_{\mathfrak{L}}\}.$$

We define the creation operators $\widetilde{S}_\beta, \beta \in \Sigma$, on $\mathfrak{H}_{\mathfrak{L}}$ by

$$\widetilde{S}_\beta e_x = \begin{cases} e_{\beta x} & \text{if } \beta \in \Gamma_1^-(x), \\ 0 & \text{if } \beta \notin \Gamma_1^-(x). \end{cases}$$

If $\mathfrak{L}$ satisfies condition (I) and is predecessor-separated, then $\mathcal{O}_{\mathfrak{L}}$ is isomorphic to the C^*-algebra $C^*(\widetilde{S}_\beta, \beta \in \Sigma)$ generated by the partial isometries $\widetilde{S}_\beta, \beta \in \Sigma$, on the Hilbert space $\mathfrak{H}_{\mathfrak{L}}$. By this observation, we know that if a λ-graph system $\mathfrak{L}$ is the canonical λ-graph system for a subshift Λ, then the C^*-algebra $\mathcal{O}_{\mathfrak{L}}$ is isomorphic to $\mathcal{O}_\Lambda$ in a natural way. If in particular $\mathfrak{L}$ comes from a finite directed graph G, the algebra $\mathcal{O}_{\mathfrak{L}}$ becomes the Cuntz-Krieger algebra $\mathcal{O}_{A_G}$ associated to its adjacency matrix A_G. Let E_G be its edge set. We indeed see that the generating partial isometries $S_e, e \in E_G$, satisfy the relations

$$\sum_{e\in E_G} S_e S_e^* = 1, \qquad S_e^* S_e = \sum_{f\in E_G} A_G(e,f) S_f S_f^*, \qquad e \in E_G.$$

Some generalizations of Cuntz-Krieger algebras from a graph point of view have been studied in [KPR], [KPRR], [Tom] (cf. [KPW2], [EL], etc.).

---

[2]The relation (5.5) is written in the original paper instead of (5.3).

We generalize irreducibility and aperiodicity for finite directed graphs to λ-graph systems.

(i) A λ-graph system $\mathfrak{L}$ is said to be *irreducible* if for a vertex $v \in V_l$ and an ι-orbit $x = (x_1, x_2, \dots) \in \Omega_{\mathfrak{L}} = \underleftarrow{\lim} V_l$, there exists a path in $\mathfrak{L}$ starting at v and terminating at x_{l+N} for some $N \in \mathbb{N}$.

(ii) A λ-graph system $\mathfrak{L}$ is said to be *aperiodic* if for a vertex $v \in V_l$ there exists an $N \in \mathbb{N}$ such that there exist paths in $\mathfrak{L}$ starting at v and terminating at all the vertices of V_{l+N}.

THEOREM 5.3 ([MA12]). *If $\mathfrak{L}$ satisfies condition* (I) *and is irreducible, then the C^*-algebra $\mathcal{O}_{\mathfrak{L}}$ is simple. In particular, if $\mathfrak{L}$ is aperiodic, then $\mathcal{O}_{\mathfrak{L}}$ is simple and purely infinite.*

If $\mathcal{O}_{\mathfrak{L}}$ is not simple, the quotient C^*-algebra by an ideal is also a C^*-algebra associated with a λ-graph system that is a sub-λ-graph system of the original λ-graph system ([Ma14]). Hence the class of the C^*-algebras associated with λ-graph systems is closed under quotients by its ideals. It is expected that the class of C^*-algebras associated with λ-graph systems contains many purely infinite simple nuclear C^*-algebras. Similarly to the cases of Cuntz-Krieger algebras and C^*-algebras associated with subshifts, there exists an action $\alpha_{\mathfrak{L}}$ of the torus group $\mathbb{T}$ on the algebra $\mathcal{O}_{\mathfrak{L}}$ that is called the gauge action. It satisfies $\alpha_{\mathfrak{L}t}(S_\alpha) = tS_\alpha, \alpha \in \Sigma$, for $t \in \mathbb{T}$. The fixed point subalgebra ${\mathcal{O}_{\mathfrak{L}}}^{\alpha_{\mathfrak{L}}}$ of $\mathcal{O}_{\mathfrak{L}}$ under $\alpha_{\mathfrak{L}}$ is an AF-algebra $\mathcal{F}_{\mathfrak{L}}$, whose embedding rule is defined by the nonnegative matrix system (M, I) for $\mathfrak{L}$. The algebra $\mathcal{F}_{\mathfrak{L}}$ is stably isomorphic to the crossed product $\mathcal{O}_{\mathfrak{L}} \rtimes_{\alpha_{\mathfrak{L}}} \mathbb{T}$.

We will state a relationship between the C^*-algebra $\mathcal{O}_{\mathfrak{L}}$ and the shift equivalence invariants: the dimension triple, the K-groups and the Bowen-Franks groups for the nonnegative matrix system (M, I) defined in Section 4. The generalization of Theorem 4.3 to not necessarily canonical symbolic matrix systems holds:

THEOREM 5.4 ([MA12]). *Let (M, I) be the nonnegative matrix system of the symbolic matrix system $(\mathcal{M}, I)$ for a λ-graph system $\mathfrak{L}$. We then have isomorphisms*

$$
\begin{aligned}
(\Delta_{(M,I)}, \Delta^+_{(M,I)}, \delta_{(M,I)}) &\cong (K_0(\mathcal{F}_{\mathfrak{L}}), K_0(\mathcal{F}_{\mathfrak{L}})_+, \hat{\alpha}_{\mathfrak{L}*}),\\
K_i(M, I) &\cong K_i(\mathcal{O}_{\mathfrak{L}}), \qquad i = 0, 1,\\
BF^i(M, I) &\cong \operatorname{Ext}^{i+1}(\mathcal{O}_{\mathfrak{L}}), \qquad i = 0, 1.
\end{aligned}
$$

The C^*-algebra $\mathcal{O}_{\mathfrak{L}}$ is nuclear and satisfies the universal coefficient theorem of K-theory proved by Rosenberg and Schochet ([RS], cf. [Bro]). By the classification theory of E. Kirchberg [Ki], N. C. Phillips [Ph] and M. Rørdam [Rø2], the isomorphism classes of these C^*-algebras for aperiodic λ-graph systems are completely classified by their K-groups. We indeed see by a result of Rørdam [Rø2, Proposition 6.7],

COROLLARY 5.5. *For an aperiodic λ-graph system $\mathfrak{L}$, the C^*-algebra $\mathcal{O}_{\mathfrak{L}}$ is an inductive limit of the sequence of simple Cuntz-Krieger algebras.*

As the algebra $\mathcal{O}_{\mathfrak{L}}$ is a groupoid C^*-algebra defined from a continuous graph, it is also realized as a C^*-algebra constructed from a Hilbert C^*-bimodule by a general theory of C^*-algebras from continuous graphs ([De3], cf. [Pi], [Kat], [KPW], [KW], [PWY]).

6. Examples

We will finally give some examples.

EXAMPLE 1. *The context free shift*: Let Z be a subshift over $\{1,2,3\}$ whose forbidden words are $\{3\underbrace{2\cdots 2}_{m}\underbrace{1\cdots 1}_{k}3 \mid m \neq k\}$. Its concatenation rule comes from a context free language. The subshift Z is named the context free shift in [LM]. It is a synchronizing countershift, and it is not sofic. We denote by (M^Z, I^Z) its canonical nonnegative matrix system. Then we have

$$K_0(M^Z, I^Z) \cong \mathbb{Z}, \quad K_1(M^Z, I^Z) \cong BF^0(M^Z, I^Z) \cong 0, \quad BF^1(M^Z, I^Z) \cong \mathbb{Z}$$

([Ma5]). The subshift Z is aperiodic in past equivalence, so that the C^*-algebra $\mathcal{O}_Z$ is simple and purely infinite. As the K_1-group is not the torsion-free part of its K_0-group, these types of K-groups and Bowen-Franks groups do not appear in Cuntz-Krieger algebras.

EXAMPLE 2. *The Dyck shift*: Let S be the subshift over the brackets $(,),[,]$, whose forbidden words are finite sequences of these brackets that do not obey standard bracket rules (cf. [AU], [Kr]). The following words, for example, are forbidden:

$$(], \quad [), \quad ([]], \quad [[]), \quad (()], \quad [()), \ldots.$$

The subshift is called the Dyck shift with four brackets, and it is not sofic. Let us denote by (M^S, I^S) its canonical nonnegative matrix system. Then we have

$$K_0(M^S, I^S) \cong \mathbb{Z}^\infty, \quad K_1(M^S, I^S) \cong BF^0(M^S, I^S) \cong 0, \quad BF^1(M^S, I^S) \cong \prod^{\infty} \mathbb{Z}$$

([Ma11]). These types of K-groups and Bowen-Franks groups do not appear in the C^*-algebras associated with sofic shifts and topological Markov shifts. For the canonical nonnegative matrix system $(M^{S\times[n]}, I^{S\times[n]})$ of the product subshift $S \times [n]$ between S and the full n-shift $[n]$, we have

$$K_0(M^{S\times[n]}, I^{S\times[n]}) \cong \mathbb{Z}[\tfrac{1}{n}]^\infty, \quad K_1(M^{S\times[n]}, I^{S\times[n]}) \cong 0,$$

$$BF^0(M^{S\times[n]}, I^{S\times[n]}) \cong \prod_{\mathbb{N}} (\varprojlim \mathbb{Z}/n^i\mathbb{Z})/\mathbb{Z} \cong \prod_{\mathbb{N}} \frac{n\text{-adic infinite polynomials}}{n\text{-adic finite polynomials}},$$

$$BF^1(M^{S\times[n]}, I^{S\times[n]}) \cong 0,$$

where $\prod_{\mathbb{N}}(\varprojlim \mathbb{Z}/n^i\mathbb{Z})/\mathbb{Z}$ is the abelian group of the countable infinite products of the quotient group of the natural projective limit $\mathbb{Z}/n\mathbb{Z} \leftarrow \mathbb{Z}/n^2\mathbb{Z} \leftarrow \cdots$ by $\mathbb{Z}$ in a natural way.

These results imply that the subshift Z of Example 1 is not flow equivalent to $S \times [n]$ for any $n = 1, 2, \cdots$. And also we know that $S \times [n]$ is not flow equivalent to $S \times [m]$ for distinct n, m.

EXAMPLE 3. *The λ-graph system from a Shannon graph*: A Shannon graph is a labeled directed graph with possibly an infinite number of vertices. It is required that the set of all labels is finite, and its labeling is left-resolving (generally right-resolving). In [KM], a method to construct a λ-graph system from a Shannon graph has been introduced in the following way. Let $\mathcal{S}$ be a Shannon graph over Σ with vertex set V. For a vertex $v \in V$, let $\Gamma_l^-(v)$ be the set of all words of length l terminating at v. Consider the equivalence relation $v \approx_{(l)} v'$ on V defined by $\Gamma_l^-(v) = \Gamma_l^-(v')$ for each l. We denote by $V_l = \{\mathfrak{v}_1^l, \dots, \mathfrak{v}_{m(l)}^l\}$ the equivalence classes of V by $\approx_{(l)}$, where V_0 means V. The set of equivalence classes $\mathfrak{v}_i^l, i = 1, \dots, m(l)$, is identified with the set $\{\Gamma_l^-(v) : v \in V\}$ of words of length l. We define a map $\iota_{l,l+1} : V_{l+1} \to V_l$ by $\iota_{l,l+1}(\mathfrak{v}_j^{l+1}) = \mathfrak{v}_i^l$ for $\mathfrak{v}_j^{l+1} \subset \mathfrak{v}_i^l$. If there exists an edge labeled $\omega \in \Sigma$ from a vertex belonging to $\mathfrak{v}_i^l$ to a vertex belonging to $\mathfrak{v}_j^{l+1}$, we define an edge from $\mathfrak{v}_i^l$ to $\mathfrak{v}_j^{l+1}$ labeled ω. These define a left-resolving λ-graph system over Σ. We denote it by $\mathfrak{L}(\mathcal{S})$. Consider the following Shannon garph $\mathcal{S}$ over $\Sigma = \{a, b, c, d, e\}$:

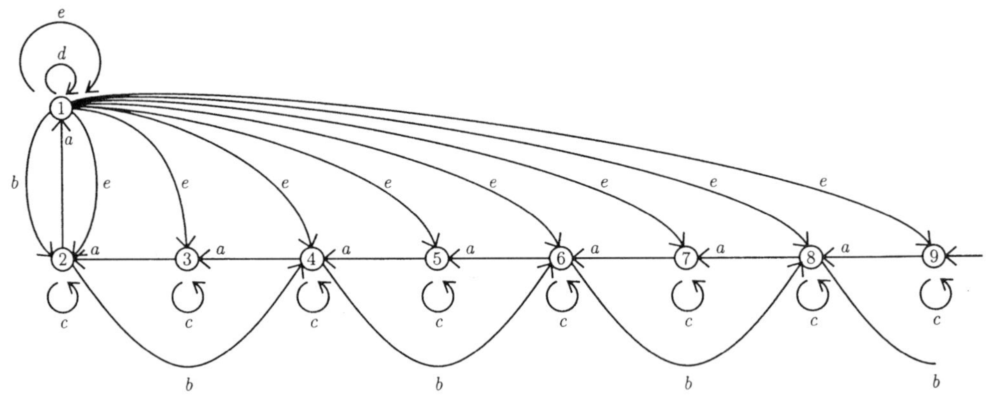

The λ-graph system $\mathfrak{L}(\mathcal{S})$ for $\mathcal{S}$ is aperiodic, so that the resulting C^*-algebra $\mathcal{O}_{\mathfrak{L}(\mathcal{S})}$ is simple and purely infinite. Its K-groups and Ext-groups are

$$K_0(\mathcal{O}_{\mathfrak{L}(\mathcal{S})}) \cong 0, \quad K_1(\mathcal{O}_{\mathfrak{L}(\mathcal{S})}) \cong \mathbb{Z}, \quad \mathrm{Ext}^0(\mathcal{O}_{\mathfrak{L}(\mathcal{S})}) \cong 0, \quad \mathrm{Ext}^1(\mathcal{O}_{\mathfrak{L}(\mathcal{S})}) \cong \mathbb{Z}$$

([Ma12]). The C^*-algebra $\mathcal{O}_{\mathfrak{L}(\mathcal{S})}$ is not stably isomorphic to any Cuntz algebra $\mathcal{O}_n, n = 2, 3, \dots, \infty$, nor to any Cuntz-Krieger algebra $\mathcal{O}_A$.

Acknowledgment

The author would like to thank Masaru Nagisa and Jun Tomiyama for their helpful suggestions. The author would also like to thank the referee for valuable comments on the original manuscript.

References

[AU] A. V. Aho and J. Ullman, *The theory of language*, Math. Systems Theory **2** (1968), 97–125.

[An] C. Anantharaman-Delaroche, *Purely infinite C^*-algebras arising from dynamical systems*, Bull. Soc. Math. France **125(2)** (1997), 199–225.

[Bl] B. Blackadar, *K-Theory for Operator Algebras*, Springer-Verlag, Berlin, Heidelberg and New York, 1986.

[BG] F. P. Boca and P. Goldstein, *Topological entropy for the canonical endomorphism of the Cuntz-Krieger algebras*, Bull. London Math. Soc. **32** (2000), 345-352.

[BF] R. Bowen and J. Franks, *Homology for zero-dimensional nonwandering sets*, Ann. Math. **106** (1977), 73–92.

[BK] M. Boyle and W. Krieger, *Almost Markov and shift equivalent sofic systems*, Proceedings of Maryland Special Year in Dynamics 1986-87, Springer -Verlag Lecture Notes in Math **1342** (1988), 33–93.

[Bra] O. Bratteli, *Inductive limits of finite-dimensional C^*-algebras*, Trans. Amer. Math. Soc. **171** (1972), 195–234.

[BraKi] O. Bratteli and A. Kishimoto, *Trace scaling automorphisms of certain stable AF algebras II*, Quart. J. Math. Oxford Ser. (2) **51** (2000), 131–154.

[Bro] L. G. Brown, *The universal coefficient theorem for Ext and quasidiagonality*, Operator Algebras and Group Representation, Monogr. Stud. Math. 17, Pitman Press, Boston, MA, 1984, pp. 60–64.

[CaE] T. M. Carlsen and S. Eilers, *Augmenting dimension group invariants for substitutional dynamical system* (preliminary version), 2002.

[CaM] T. M. Carlsen and K. Matsumoto, *Some remarks on the C^*-algebras associated with subshifts*, preprint, 2001.

[Ch] M. Choda, *Endomorphisms of shift type (entropy for endomorphisms of Cuntz algebras)*, In : Operator Algebras and Quantum Field Theory, Proceedings, Rome 1996, International Press, Cambridge, MA, 1997, pp.469-475.

[C] J. Cuntz, *Simple C^*-algebras generated by isometries*, Commun. Math. Phys. **57** (1977), 173–185.

[C2] J. Cuntz, *K-theory for certain C^*-algebras*, Ann. Math. **113** (1981), 181–197.

[C3] J. Cuntz, *A class of C^*-algebras and topological Markov chains II: reducible chains and the Ext-functor for C^*-algebras*, Invent. Math. **63** (1981), 25–40.

[CK] J. Cuntz and W. Krieger, *A class of C^*-algebras and topological Markov chains*, Invent. Math. **56** (1980), 251–268.

[Da] K. Davidson, *C^*-algebras by example*, Fields Institute Monographs 6, Amer. Math. Soc., Providence, Rhode Island, 1996.

[De] V. Deaconu, *Groupoids associated with endomorphisms*, Trans. Amer. Math. Soc. **347** (1995), 1779–1786.

[De2] V. Deaconu, *Generalized Cuntz-Krieger algebras*, Proc. Amer. Math. Soc. **124** (1996), 3427–3435.

[De3] V. Deaconu, *Generalized solenoids and C^*-algebras*, Pacific J. Math. **190** (1999), 247–260.

[De4] V. Deaconu, *Continuous graphs and C^*-algebras*, Operator Theoretical Methods (Timişoara, 1998) Theta Found., Bucharest (2000), 137–149.

[Ef] E. G. Effros, *Dimensions and C^*-algebras*, CBMS Reginal Conference Series in Mathematics, vol 46, Amer. Math. Soc., Providence, RI, 1981.

[EFW] M. Enomoto, M. Fujii and Y. Watatani, *Automorphisms on Cuntz algebras*, Math. Japon. **24** (1979), 231–234.

[EFW2] M. Enomoto, M. Fujii and Y. Watatani, *K_0-groups and classification of Cuntz-Krieger algebras*, Math. Japon. **26** (1981), 443–460.

[EFW3] M. Enomoto, M. Fujii and Y. Watatani, *KMS states for gauge action on $\mathcal{O}_A$*, Math. Japon. **29** (1984), 607–619.

[EK] D. Evans and Y. Kawahigashi, *Quantum Symmetries on Operator Algebras*, Oxford University Press, Oxford, 1997.

[EL] R. Excel and M. Laca, *Cuntz-Krieger algebras for infinite matrices*, J. Reine. Angew. Math. **512** (1999), 119–172.

[Fi] R. Fischer, *Sofic systems and graphs*, Monatsh. Math. **80** (1975), 179–186.

[Fr] J. Franks, *Flow equivalence of subshifts of finite type*, Ergodic Theory Dynam. Systems **4** (1984), 53–66.

[Gil] P. M. Gilmer, *Topological quantum field theory and strong shift equivalence*, Canad. Math. Bull. **42** (1999), 190–197.

[GPS] T. Giordano, I. F. Putnam and C. F. Skau, *Topological orbit equivalence and C^*-crossed products*, J. Reine Angew. Math. **469** (1995), 51–111.

[Had] J. Hadamard, *Les surfaces à courbures opposées et leurs lignes géodésiques*, J. Math. Pures Appl. **4** (1898), 27–73.

[HN] T. Hamachi and M. Nasu, *Topological conjugacy for 1-block factor maps of subshifts and sofic covers*, Proceedings of Maryland Special Year in Dynamics 1986-87, Springer Lecture Notes in Math **1342** (1988), 251–260.

[HPS] R. H. Hermann, I. F. Putnam and C. F. Skau, *Ordered Bratteli diagrams, dimension groups and topological dynamics*, Internat. J. Math. **3** (1992), 827–864.

[H] D. Huang, *Flow equivalence of reducible shifts of finite type*, Ergodic Theory Dynam. Systems **14** (1994), 695–720.

[H2] D. Huang, *Flow equivalence of reducible shifts of finite type and Cuntz-Krieger algebras*, J. Reine Angew. Math. **462** (1995), 185–217.

[IH] Y. Itoh and T. Hamachi, *Ergodic Theory and von Neumann Algebras (Japanese)*, Kinokuniya, Tokyo, 1992.

[IT] S. Ito and Y. Takahashi, *Markov subshifts and realization of β-expansions*, J. Math. Soc. Japan **26** (1974), 33–55.

[KPW] T. Kajiwara, C. Pinzari and Y. Watatani, *Ideal structure and simplicity of the C^*-algebras generated by Hilbert modules*, J. Funct. Anal. **159** (1998), 295–322.

[KPW2] T. Kajiwara, C. Pinzari and Y. Watatani, *Hilbert C^*-bimodules and countably generated Cuntz-Krieger algebras*, J. Operator Theory **45** (2001), 3–18.

[KW] T. Kajiwara and Y. Watatani, *Hilbert C^*-bimodules and continuous Cuntz-Krieger algebras*, J. Math. Soc. Japan **54** (2002), 35–59.

[Kat] Y. Katayama, *Generalized Cuntz algebras $\mathcal{O}_N^M$*, RIMS kokyuroku **858** (1994), 87–90.

[KT] Y. Katayama and H. Takehana, *On automorphisms of generalized Cuntz algebras*, Internat. J. Math. **9** (1998), 493–512.

[KMW] Y. Katayama, K. Matsumoto and Y. Watatani, *Simple C^*-algebras arising from β-expansion of real numbers*, Ergodic Theory Dynam. Systems **18** (1998), 937–962.

[[KeP]] D. Kerr and C. Pinzari, *Noncommutative pressure and the variational principle in Cuntz-Krieger-type C^*-algebras*, J. Funct. Anal. **188** (2002), 156–215.

[KimR] K. H. Kim and F. W. Roush, *Williams conjecture is false for irreducible subshifts*, Ann. Math. **149** (1999), 545–558.

[Kir] E. Kirchberg, *The classification of purely infinite C^*-algebras using Kasparov's theory*, preprint,1994.

[Kit] B. P. Kitchens, *Symbolic Dynamics*, Springer-Verlag, Berlin, Heidelberg and New York, 1998.

[Kr] W. Krieger, *On the uniqueness of the equilibrium state*, Math. Systems Theory **8** (1974), 97–104.

[Kr2] W. Krieger, *On dimension functions and topological Markov chains*, Invent. Math. **56** (1980), 239–250.

[Kr3] W. Krieger, *On dimension for a class of homeomorphism groups*, Math. Ann. **252** (1980), 87–95.

[Kr4] W. Krieger, *On sofic systems I*, Israel J. Math. **48** (1984), 305–330.

[Kr5] W. Krieger, *On sofic systems II*, Israel J. Math. **60** (1987), 167–176.

[KM] W. Krieger and K. Matsumoto, *Shannon graphs, subshifts and lambda-graph systems*, to appear in J. Math. Soc. Japan.

[KPR] A. Kumjian, D. Pask and I. Raeburn, *Cuntz-Krieger algebras of directed graphs*, Pacific J. Math. **184** (1998), 161–174.

[KPRR] A. Kumjian, D. Pask, I. Raeburn and J. Renault, *Graphs, groupoids and Cuntz-Krieger algebras*, J. Funct. Anal. **144** (1997), 505–541.

[LM] D. Lind and B. Marcus, *An introduction to symbolic dynamics and coding*, Cambridge University Press, 1995.

[Ma] K. Matsumoto, *On C^*-algebras associated with subshifts*, Internat. J. Math. **8** (1997), 357–374.

[Ma2] K. Matsumoto, *K-theory for C^*-algebras associated with subshifts*, Math. Scand. **82** (1998), 237–255.

[Ma3] K. Matsumoto, *Dimension groups for subshifts and simplicity of the associated C^*-algebras*, J. Math. Soc. Japan **51** (1999), 679-698.

[Ma4] K. Matsumoto, *Relations among generators of C^*-algebras associated with subshifts*, Internat. J. Math. **10** (1999), 385–405.

[Ma5] K. Matsumoto, *A simple C^*-algebra arising from a certain subshift*, J. Operator Theory **42** (1999), 351-370.

[Ma6] K. Matsumoto, *Presentations of subshifts and their topological conjugacy invariants*, Documenta Math. **4** (1999), 285-340.

[Ma7] K. Matsumoto, *On automorphisms of C^*-algebras associated with subshifts*, J. Operator Theory **44** (2000), 91-112.

[Ma8] K. Matsumoto, *Stabilized C^*-algebras constructed from symbolic dynamical systems*, Ergodic Theory Dynam. Systems **20** (2000), 821-841.

[Ma9] K. Matsumoto, *Bowen-Franks groups as an invariant for flow equivalence of subshifts*, Ergodic Theory Dynam. Systems **21** (2001), 1831-1842.

[Ma10] K. Matsumoto, *Bowen-Franks groups for subshifts and Ext-groups for C^*-algebras*, K-Theory **23** (2001), 67-104.

[Ma11] K. Matsumoto, *K-theoretic invariants and conformal measures of the Dyck subshifts*, preprint, 1999.

[Ma12] K. Matsumoto, *C^*-algebras associated with presentations of subshifts*, Documenta Math. **7** (2002), 1–30.

[Ma13] K. Matsumoto, *Strong shift equivalence of symbolic dynamical systems and Morita equivalence of C^*-algebras*, preprint, 2001.

[Ma14] K. Matsumoto, *C^*-algebras associated with presentations of subshifts II. ideal structure and λ-graph subsystems*, preprint, 2001.

[Ma15] K. Matsumoto, *On strong shift equivalence of symbolic dynamical systems,* to appear in Ergodic Theory Dynam. Systems.

[MWY] K. Matsumoto, Y. Watatani and M. Yoshida, *KMS-states for gauge actions on C^*-algebras associated with subshifts*, Math. Z. **228** (1998), 489–509.

[Mo] J. Monforte, *The digital reproduction of sound*, Scientific American **Dec** (1984), 78–84.

[N] M. Nasu, *Topological conjugacy for sofic shifts*, Ergodic Theory Dynam. Systems **6** (1986), 265–280.

[N2] M. Nasu, *Textile systems for endomorphisms and automorphisms of the shift*, Mem. Amer. Math. Soc. **546** (1995).

[Pa] W. Parry, *On the β-expansion of real numbers*, Acta Math. Acad. Sci. Hungar. **11** (1960), 401–416.

[Pa2] W. Parry, *Intrinsic Markov chains*, Trans. Amer. Math. Soc. **112** (1964), 55–66.

[PS] W. Parry and D. Sullivan, *A topological invariant for flows on one-dimensional spaces*, Topology **14** (1975), 297–299.

[Ph] N. C. Phillips, *A classification theorem for nuclear purely infinite simple C^*-algebras*, Documenta Math. **5** (2000), 49–114.

[Pi] M. V. Pimsner, *A class of C^*-algebras generalizing both Cuntz-Krieger algebras and crossed product by $\mathbb{Z}$*, in Free Probability Theory, Fields Institute Communications **12**, Amer. Math. Soc., Providence, RI, 1996, 189–212.

[PWY] C. Pinzari, Y. Watatani and K. Yonetani, *KMS states, entropy and the variational principle in full C^*-dynamical systems*, Commun. Math. Phys. **213** (2000), 331–381.

[Re] J. N. Renault, *A groupoid approach to C^*-algebras*, Springer Lecture Notes in Math. **793** (1980).

[Ré] A. Rényi, *Representations for real numbers and their ergodic properties*, Acta Math. Acad. Sci. Hungar. **8** (1957), 477–493.

[Rø] M. Rørdam, *Classification of Cuntz-Krieger algebras*, K-Theory **9** (1995), 31–58.

[Rø2] M. Rørdam, *Classification of certain infinite simple C^*-algebras*, J. Funct. Anal. **131** (1995), 415–458.

[RS] J. Rosenberg and C. Schochet, *The Künneth theorem and the universal coefficient theorem for Kasparov's generalized K-functor*, Duke Math. J. **55** (1987), 431–474.

[SW] D. S. Silver and S. G. Williams, *Knot invariants from symbolic dynamical systems*, Trans. Amer. Math. Soc. **351** (1999), 3243–3265.

[Tom] M. Tomforde, *C^*-algebras of labeled graphs*, preprint.

[To] J. Tomiyama, *Invitation to C^*-algebras and topological dynamics*, World Scientific, Singapore, 1987.

[To2] J. Tomiyama, *The interplay between topological dynamics and theory of C^*-algebras*, Lecture Note No 2, Res. Inst. Math. Seoul, 1992.

[To3] J. Tomiyama, *Topological full groups and structure of normalizers in transformation group C^*-algebras*, Pacific. J. Math. **173** (1996), 571–583.

[Wa] J. B. Wagoner, *Topological Markov chains, C^*-algebras and K_2*, Advances in Math. **71** (1988), 133–185.

[We] B. Weiss, *Subshifts of finite type and sofic systems*, Monatsh. Math. **77** (1973), 462–474.

[Wi] R. F. Williams, *Classification of subshifts of finite type*, Ann. Math. **98** (1973), 120–153, erratum, Ann. Math. 99(1974), 380 − 381.

Department of Mathematical Sciences, Yokohama City University, Seto, Kanazawa-ku, Yokohama 236-0027, Japan
E-mail address: kengo@yokohama-cu.ac.jp

Translated by KENGO MATSUMOTO

Amer. Math. Soc. Transl.
(2) Vol. **211**, 2003

Inverse Problems for Elasticity

Gen Nakamura

In this paper, I will consider inverse problems for elasticity, with and without residual stress. The inverse problem for elasticity with residual stress may not be such a familiar subject; but it is also very interesting and important in applications. Henceforth we will call the inverse problems for both of these equations inverse problems for elasticity.

One purpose of this paper is to provide some knowledge and an orientation about the inverse problems for elasticity. Therefore, I will formulate the inverse problems for elasticity in a general form as much as possible. Another purpose of this paper is to review the known results, including my own. Of course, it is difficult to give a complete review of the known results. The number of researchers in this field is still small, and there are not enough results. So, I will also point out some of the open problems and conjecture possible future results.

1. Linear elastic equation

Let us consider an infinitesimally small deformation $\Omega \ni x \longmapsto X(x) \in X(\Omega)$ of a domain $\Omega \subset R^3$. We assume that the map X which describes this deformation is bijective, and identify Ω and $X(\Omega)$. We call this Ω a linear elastic medium if we have the following Hooke's law (i.e. the stress-strain relation) between the infinitesimally small strain, called linear strain, caused by this deformation and the associated stress, which is a force resisting the deformation through the surface of any subdomain of Ω:

$$\sigma_{ij} = \sum_{k,l=1}^{3} C_{ijkl}\varepsilon_{kl} \quad (1 \le i, j \le 3) \tag{1.1}$$

at each point $x = (x_1, x_2, x_3) \in \Omega$, where $\sigma = \sigma(x) = (\sigma_{ij})_{1\le i,j\le 3}$, $\varepsilon = \varepsilon(x) = (\varepsilon_{ij})_{1\le i,j\le 3}$ are the stress tensor and linear strain tensor, respectively. The linear strain tensor $\varepsilon_{ij}(x)$ is given in terms of the displacement vector $u(x) = (u_1, u_2, u_3) = X(x) - x$ at x by the following formula:

$$\varepsilon_{ij}(x) = \frac{1}{2}\left(\frac{\partial u_i}{\partial x_j} + \frac{\partial u_j}{\partial x_i}\right). \tag{1.2}$$

This article originally appeared in Japanese in Sūgaku **53** (2) (2001), 113–124. This was originally a lecture given at Kyoto University on September 25, 2000.

2000 *Mathematics Subject Classification.* Primary 35R30; Secondary 35-02, 35Q72, 74B05, 74G75.

$C = C(x) = (C_{ijkl})_{1\leq i,j,k,l\leq 3}$ is called the elasticity tensor. By the equilibrium of the angular moment, σ satisfies

(1.3) $$\sigma_{ij} = \sigma_{ji} \quad (1 \leq i, j \leq 3)$$

at each point $x \in \Omega$. Combining this with the symmetry of the ε_{ij} (i.e. $\varepsilon_{ij} = \varepsilon_{ji} \quad (1 \leq i, j \leq 3)$), we have

(1.4) $$C_{ijkl} = C_{jikl} = C_{ijlk} \quad (1 \leq i, j, k, l \leq 3)$$

at each point $x \in \Omega$. Also, due to the fact that $C_{ijkl} = \dfrac{\partial^2 W}{\partial\varepsilon_{kl}\partial\varepsilon_{ij}}(\varepsilon)$ is given in terms of the strain energy function $W(\varepsilon) = \dfrac{1}{2}\sum_{i,j,k,l=1}^{3} C_{ijkl}\varepsilon_{ij}\varepsilon_{kl}$, we have

(1.5) $$C_{ijkl} = C_{klij} \quad (1 \leq i, j, k, l \leq 3)$$

at each point $x \in \Omega$, by changing the order of differentiation.

Henceforth, we assume the strong convexity condition for $W(\varepsilon)$. Namely, there exists $\delta > 0$ such that

(1.6) $$W(\varepsilon) \geq \delta \sum_{i,j=1}^{3} \varepsilon_{ij}^2$$

for any $x \in \bar{\Omega}$ and symmetric matrix $\varepsilon = (\varepsilon_{ij})$. This means that any nonzero linear strain produces a positive strain energy.

If a body force $\mathrm{b} = \mathrm{b}(x) = (\mathrm{b}_1, \mathrm{b}_2, \mathrm{b}_3)$ acts on a linear elastic medium Ω with density $\rho(x) > 0$, the equation for the equilibrium of forces acting on Ω is

(1.7) $$\sum_{j=1}^{3} \frac{\partial}{\partial x_j}\sigma_{ij} + \rho(x)\mathrm{b}_i(x) = 0 \quad (x \in \Omega, 1 \leq i \leq 3).$$

This equation is derived by first considering the equilibrium of forces acting on an arbitrarily small portion ω of Ω and transforming the stress $\int_{\partial\omega} \sigma_{ij} n_j dS$ acting on ω through its surface to a volume integral by applying Gauss's divergence theorem, where dS denotes the surface element of $\partial\omega$. Then, recalling that ω is arbitrary, we have the differential equation (1.7). Here, since the stress resists the deformation, note that the direction of the unit normal vector $n = (n_1, n_2, n_3)$ of $\partial\omega$ is pointed outside ω.

Now, if we consider an infinitesimal deformation of Ω depending on time, the equation of motion of Ω is

(1.8)
$$\rho(x)\frac{\partial^2 u_i}{\partial t^2}(t, x) = \sum_{j=1}^{3} \frac{\partial}{\partial x_j}\sigma_{ij}(t, x) + \rho(x)\mathrm{b}_i(t, x) \quad (x \in \Omega, 0 < t < T, 1 \leq i \leq 3).$$

Here t is the time, and we note that the displacement u, linear strain tensor ε, stress tensor σ, and body force b all generally depend on time, while the density ρ and elasticity tensor C do not. We call (1.7) and (1.8) a static elastic equation and dynamic elastic equation, respectively.

The elastic property of an elastic medium Ω is determined by its elastic tensor C. If C does not depend on the direction, we call C isotropic, and otherwise anisotropic. Also, if C does not depend on the position, we call C homogeneous, and otherwise inhomogeneous. For each isotropic tensor C, it is known that there exist two functions $\lambda(x), \mu(x)$, called Lamé coefficients, such that

$$C_{ijkl}(x) = \lambda(x)\delta_{ij}\delta_{kl} + \mu(x)(\delta_{ik}\delta_{jl} + \delta_{il}\delta_{kj}) \quad (x \in \Omega, 1 \leq i, j, k, l \leq 3), \tag{1.9}$$

where δ_{ij} is Kronecker's delta ([**8**]). In this case, (1.6) is equivalent to

$$\mu(x) > 0, \ 3\lambda(x) + 2\mu(x) > 0 \quad (x \in \bar{\Omega}) \tag{1.10}$$

(see [**22**]).

2. Elastic equation with residual stress

Linear elastic media with residual stress are very interesting to study together with linear elastic media. In this paper, for simplicity we consider an isotropic elastic medium with residual stress. Suppose an isotropic medium with Lamé coefficients $\lambda(x), \mu(x)$ without any stress inside (N.B.: such a state is called natural) has been deformed and came to an equilibrium state (called the first equilibrium state). We denote the region occupied by this isotropic medium in the first equilibrium state by Ω and denote its stress tensor at each point $x = (x_1, x_2, x_3)$ by $T = T(x) = (T_{ij})_{1 \leq i,j \leq 3}$. This T is called the residual stress. T itself satisfies the equations of equilibrium of force in Ω and on $\partial\Omega$:

$$\sum_{j=1}^{3} \frac{\partial}{\partial x_j} T_{ij} + \rho \mathrm{b}_i = 0 \quad \text{in } \Omega \quad (1 \leq i \leq 3), \tag{2.1}$$

$$\sum_{j=1}^{3} T_{ij} n_j = g_i \quad \text{on } \partial\Omega \quad (1 \leq i \leq 3), \tag{2.2}$$

and we can take $g = (g_1, g_2, g_3) = 0$ if $\mathrm{b} = 0$. Here $n = (n_1, n_2, n_3)$ is the unit outer normal vector of $\partial\Omega$ and $-g$ is the boundary traction necessary to support the traction created by the residual stress at the boundary $\partial\Omega$. Henceforth, the direction of the outer normal vector of the boundary of a bounded domain is directed outside this bounded domain. We note that by the equilibrium of the angular momentum or the fact that residuat stress itself is a stress, T is symmetric.

Next we deform this Ω, which is in the first equilibrium state, by an infinitesimal deformation, and as the result of this deformation the deformed Ω is in the second equilibrium state, or a state of motion. By using the theory of nonlinear elasticity and neglecting those terms whose smallness is of degree higher than 1, the equilibrium of force and the equation of motion have the same forms as (1.7) and (1.8), respectively. But we have to replace σ by

$$\begin{aligned} \sigma &= \left(\sigma_{ij} = \sum_{k,l=1}^{3} L_{ijkl} \frac{\partial u_k}{\partial x_l} \right)_{1 \leq i,j \leq 3} \\ &= T + HT + \lambda(\mathrm{tr}\varepsilon)I + 2\mu\varepsilon + \beta_1(\mathrm{tr}\varepsilon)(\mathrm{tr}T)I + \beta_2(\mathrm{tr}T)\varepsilon \\ &\quad + \beta_3((\mathrm{tr}\varepsilon)T + (\mathrm{tr}(\varepsilon T))I) + \beta_4(\varepsilon T + T\varepsilon) \end{aligned} \tag{2.3}$$

(see [**32**]), where I is the identity matrix, $u = u(x) = (u_1, u_2, u_3)$ is the displacement vector of the deformation of Ω, $\mathrm{tr}\ T = \sum_{i=1}^{3} T_{ii}$, and the β_i $(1 \leq i \leq 4)$ are physical coefficients determined from the Lamé coefficients and the elasticity coefficients of order 3 of our isotropic medium; the β_i can be assumed very small. The elasticity coefficients of order 3 are the third order derivatives of the stored energy function $w(E)$ at $E = 0$ given in terms of the second Piola-Kirchhoff stress and strain $E = \varepsilon + \frac{1}{2}{}^t(\nabla u)\,\nabla u$. In linear elasticity the stored energy function $w(E)$ and the strain E are the strain energy function $W(\varepsilon)$ and linear strain ε, respectively.

We explained that the residual stress is produced by the deformation that deformed the natural state to the first equilibrium state. However, some readers might have difficulty understanding the way that the residual stress was explained here. This is because we do not know whether the natural state really exists or not, and most elastic materials have undergone some manufacturing process that involves an inelastic or plastic deformation. An accurate uniaxial tension test tells us that the familiar Hooke's law is not correct. A more correct stress-strain relation is the Hartig law for the uniaxial tension test. The multidimensional version of this law coincides with (2.3). Having this in mind, C.-S. Man defined the residual stress T as the one included in (2.3) and satisfying (2.1), (2.2) ([**20**]). The advantage of Man's definition of the residual stress based on the experimental law is that it does not require any history which led to the first equilibrium state and is consistent with the old definition of the residual stress, which I explained here based on the deformation of the natural state.

If we express (1.7) and (1.8) in terms of u using (2.3), these equations become

$$\sum_{j,k,l=1}^{3} \frac{\partial}{\partial x_j}\Big(L_{ijkl}\frac{\partial u_k}{\partial x_l}\Big) + \rho \mathrm{b}_i = 0 \quad \text{in } \Omega \quad (1 \le i \le 3), \tag{2.4}$$

$$\rho\frac{\partial^2 u_i}{\partial t^2} = \sum_{j,k,l=1}^{3} \frac{\partial}{\partial x_j}\Big(L_{ijkl}\frac{\partial u_k}{\partial x_l}\Big) + \rho \mathrm{b}_i \quad \text{in } (0,T)\times\Omega \quad (1 \le i \le 3). \tag{2.5}$$

Moreover, L_{ijkl} satisfies only one symmetry: $L_{ijkl} = L_{klij}$ $(1 \le i,j,k,l \le 3)$ and we can assume that $L_{ijkl} - \{\lambda\delta_{ij}\delta_{kl} + \mu(\delta_{ik}\delta_{jl} + \delta_{il}\delta_{jk})\}$ $(1 \le i,j,k,l \le 3)$ are very small.

Henceforth we call (2.4) and (2.5) the static elastic equation with residual stress and the dynamical elastic equation with residual stress, respectively.

3. Direct and inverse problems for elastic equation

Let $\Omega \subset R^3$ be a domain given in §§1 and 2, which is either a linear elastic medium or an isotropic elastic medium with residual stress. For $u = (u_1, u_2, u_3)$, define the ith component $(Lu)_i$ of Lu by

$$\begin{cases} (Lu)_i = \displaystyle\sum_{j,k,l=1}^{3} \frac{\partial}{\partial x_j}\left(C_{ijkl}\frac{\partial u_k}{\partial x_l}\right) = \sum_{j=1}^{3}\frac{\partial}{\partial x_j}\sigma_{ij} \\ \left(\text{resp. } (Lu)_i = \displaystyle\sum_{j,k,l=1}^{3} \frac{\partial}{\partial x_j}\left(L_{ijkl}\frac{\partial u_k}{\partial x_l}\right) = \sum_{j=1}^{3}\frac{\partial}{\partial x_j}\sigma_{ij}\right), \end{cases} \tag{3.1}$$

where σ is given by (1.1) (resp. (2.3)). Henceforth, since we want to deal with linear elastic media and isotropic elastic media with residual stress at the same time, (resp. $\cdots$) is the description for the isotropic elastic medium with residual stress.

Let $\partial\Omega$ be the boundary of Ω, and let it be given by

$$\partial\Omega = \overline{\Gamma}_1 \cup \overline{\Gamma}_2 \cup \overline{\gamma}, \tag{3.2}$$

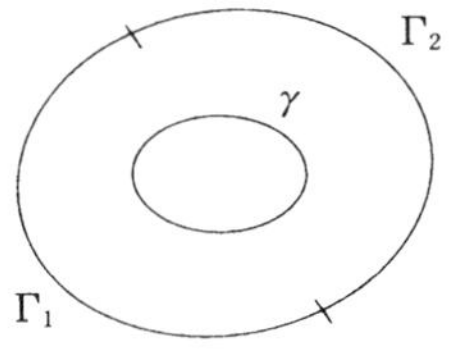

Figure 1

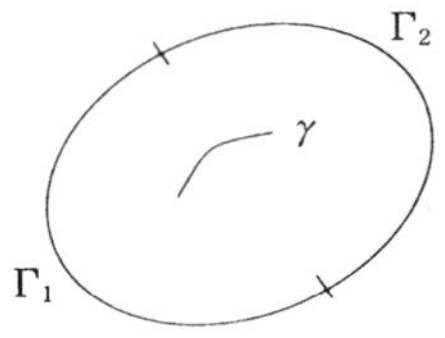

Figure 2

where γ and the Γ_i $(i = 1, 2)$ are mutually disjoint open subsets of $\partial\Omega$. Note that we do not exclude the case that some of Γ_1, Γ_2, γ are empty sets.

The typical two-dimensional figures of Ω are shown in Figures 1 and 2. In these figures, γ describes the boundary of a cavity and the shape of a crack, respectively. For the static elastic equation including the steady state case,

$$(L + \rho\omega^2)u + \rho \mathrm{b} = 0 \quad \text{in } \Omega, \tag{3.3}$$

and the dynamical elastic equation,

$$\rho \frac{\partial^2 u}{\partial t^2} = Lu + \rho \mathrm{b} \quad \text{in } (0, T) \times \Omega, \tag{3.4}$$

we consider the boundary condition on $\partial\Omega$:

$$\begin{cases} B_j u = g_j & \text{on } \Gamma_j \quad (\text{resp. } (0, T) \times \Gamma_j) \quad (j = 1, 2), \\ Du = h & \text{on } \gamma \quad (\text{resp. } (0, T) \times \gamma), \end{cases} \tag{3.5}$$

where ω denotes the frequency. Also, for the dynamical elastic equation with or without residual stress, we consider the initial condition

$$u|_{t=0} = \varphi_0, \; \left.\frac{\partial u}{\partial t}\right|_{t=0} = \varphi_1 \quad \text{in } \Omega. \tag{3.6}$$

B_j $(j = 1, 2)$ and D in (3.5) are partial differential operators on Γ_j $(j = 1, 2)$ and γ, respectively. For simplicity, we assume that these partial differential operators do not depend on time, even for the dynamical elastic equation.

If we know $L, B_j, D, \Gamma_k, \gamma, \mathrm{b}, g_l, h$ ((resp. $\rho, L, B_j, D, \Gamma_k, \gamma, \mathrm{b}, g_l, h, \varphi_m$)), the boundary value problem ((resp. initial boundary value problem)) is to obtain a solution $u = u(\mathrm{b}, g_l, h)$ ((resp. $u(\mathrm{b}, g_l, h, \varphi_m)$) that satisfies the equation and boundary condition ((resp. and also the initial condition)). Here, we have used ((resp. $\cdots$)) to deal with the static elastic equation and dynamical elastic equation at the same time, as we did for the linear elastic equation and isotropic elastic equation with residual stress.

We call those of the items in b, g_l, h ((resp. $\mathrm{b}, g_l, h, \varphi_m$)) at our disposal the input data, and we call the information we can get about $u = u(\mathrm{b}, g_l, h)$ ((resp. $u, \mathrm{b}, g_l, h, \varphi_m$))) on some part of $\overline{\Omega}$ ((resp. $[0, T] \times \overline{\Omega}$)) the output data. Suppose we do not know some of $L, B_j, D, \Gamma_k, \gamma, \mathrm{b}, g_l, h$ ((resp. $\rho, L, B_j, D, \Gamma_k, \gamma, \mathrm{b}, g_l, h, \varphi_m$)) (we call these the unknowns); then the inverse problem is to identify the unknowns from the input and output data. We assume that the boundary value problem ((resp. initial boundary value problem)) is well posed whenever we consider inverse problems, and we call the boundary value problem ((resp. initial boundary value problem)) the forward problem.

We have formulated the inverse problems for elastic equations in a general form as much as possible, but we have to examine the value of each inverse problem from its applied point of view.

In order to have a better understanding about $\Gamma_1, \Gamma_2, \gamma$, we give 4 examples.

EXAMPLE 1 (an example of the Cauchy problem). Let Ω be a bounded domain, $\gamma = \phi$, $\omega = 0$, and assume that the boundary data g_2 on Γ_2 of the boundary value problem (3.3) for the homogeneous equation is the only unknown. Suppose there exists a solution $u = u(\mathrm{b}, g_l)$ of this boundary value problem even if g_2 is unknown. This assumption of the existence of a solution is not unnatural at all when we do any kind of measurement for the unknown. Then, identify the unknown g_2 from the input data $B_1 u|_{\Gamma_1} = g_1$ and the output data $B_1' u|_{\Gamma_1} = g_1'$. We have assumed here that B_1' is defined in such a way that $B_1 u|_{\Gamma_1}$ and $B_1' u|_{\Gamma_1}$ determine the Cauchy data $(u|_{\Gamma_1}, \left.\frac{\partial u}{\partial n}\right|_{\Gamma_1})$ on Γ_1, where n is the unit outer normal vector of Γ_1. For practical applications we can only have the input and output data on a part of Γ_1. The necessity of this kind of localization of the input and output data for practical applications should always be considered for other inverse problems. We want the readers to keep this in mind.

EXAMPLE 2 (inverse problems for identifying a cavity or crack). (i) Let B_1' be as in Example 1. Also, let γ be a closed surface, $\mathrm{b} = h = 0$, $\omega = 0$ and $Du = \sigma n|_{\gamma}$, where n is the outer unit normal vector of γ directed into Ω. Suppose γ is the only unknown in the boundary value problem (3.3) for the homogeneous linear elastic equation. Then, identify the shape of the boundary γ of the cavity from the input and output data: $\{(B_1 u|_{\Gamma_1} = g_1, B_1' u|_{\Gamma_1} = g_1'); \text{finite number of } g_1\}$. Note here that g_2 in $B_2 u|_{\Gamma_2} = g_2$ can be fixed or can take several g_2 when we consider a finite number of g_1. Also, the input data and output data do not have to be taken on the same Γ_1. These facts should be taken into account in the next example (ii), and in Example 3.
(ii) By considering a crack as a retraction of a cavity, we can formulate an inverse problem analogous to (i).

EXAMPLE 3 (inverse problems for identifying the source term). Let $\gamma = \emptyset$ and assume that the source term ρb is the only unknown in the boundary value problem (resp. initial boundary value problem) given by (3.3) (resp. (3.4)). Then, identify the body force b knowing ρ from the finite number of input and output data: $\{(B_1 u|_{\Gamma_1} = g_1, B_1' u|_{\Gamma_1} = g_1'); \text{finite number of } g_1\}$. The typical example of an inverse problem for identifying the source term for the dynamical elastic equation (3.4), which clearly tells its physical and engineering meaning, is the inverse problem for identifying a seismic source. This inverse problem is to identify the place, strength and time of an earthquake by observing the seismic wave arriving at the surface of the earth.

For the static elastic equation (3.3) with elasticity tensor (C_{ijkl}), replace $\rho\mathrm{b}$ by div σ^* with $\sigma^* = (\sigma_{ij}^*) = (\sum_{k,l=1}^3 C_{ijkl}\epsilon_{kl}^*)$ and assume that the inelastic strain (ϵ_{kl}^*) is the only unknown. Then, we can consider an inverse problem for identifying (ϵ_{kl}^*) from the input and output data, which is analogous to the previous inverse problem for identifying the body force b. The inelastic strain (ϵ_{kl}^*) inside the elastic medium can be created by some damage that happened to the elastic body. So this inverse problem is related to evaluating damage.

EXAMPLE 4 (inverse problem for identifying an inclusion). Let Ω be a bounded domain, $\gamma = \emptyset$, $\mathrm{b} = \varphi_0 = \varphi_2 = 1$. Assume that the coefficients of the dynamical elastic equation (3.3) are different inside and outside some subdomain D of Ω. Then, by assuming the coefficients inside D are unknown, identify these unknown coefficients in D from the input and output data, as in (i) of Example 2.

4. Fundamental problems for the inverse problem

For simplicity, from here on, we use the word observation data for the input and output data. The real data always contain some error due to the measurement and roundoff error in their numerical realization. We distinguish the observation data without error and with some possible error by calling them genuine observation data and observation data, respectively. Although infinitely many measurements are unrealistic, we admit infinitely many measurements in our mathematical study of inverse problems. This is because an analysis using infinitely many measurements without error as observation data sometimes provides an approximate indentification of the unknown by sufficiently many finite measurements with some possible error ([**4**]). Therefore, it is not meaningless even from the applied point of view to extend the number of measurement to infinity.

Not only for the inverse problems for the elastic equation but also for those for any other equation, the fundamental problems for the mathematical study of inverse problems are the following 4 problems.

(i) *Uniqueness*: This problem asks about the uniqueness of identifying the unknown from the genuine observation data. In other words, it asks about the one-to-one correspondence between the genuine obsevation data and the unknown.

(ii) *Stability*: This problem asks about the continuous dependence of the variation of the unknown on the variation of the genuine observation data. Due to an apriori bound for the unknown and a bound for the genuine observation data with respect to some norms of the spaces in which they are included, we can use weaker norms for the estimate of variation of the unknown and genuine observation data if these norms have the interpolation property. For the stability, it does not matter if we use a priori information about the unknown based on some experience. If we denote the suprema of the variation of the genuine data and unknown by ε and $\delta(\varepsilon)$, the speed of the convergence of $\delta(\varepsilon) \longrightarrow 0$ $(\varepsilon \longrightarrow 0)$ is usually much slower than ε. Hence, a very small error for the observation datum can cause an error for the unknown which cannot be neglected. This is the typical ill-posedness that we have for the inverse problems.

(iii) *Reconstruction*: This problem is nothing but the problem of reconstructing the unknown from the genuine observation data. For applications, it is important to reconstruct the unknown approximately from the observation data.

(iv) *Existence*: This problem asks about giving a criterion for the genuine observation datum.

REMARK. Of course (iv) is a very difficult problem. Even when we can solve the existence problem, we can easily imagine that it is very difficult to verify the criterion. If the observation data have some error in (iii), there are methods, known as Moore-Penrose generalized inversion and Tikhonov's regularization method ([**39**]).

The problem of error in the observation data and ill-posedness of the inverse problem are big problems for inverse problems. Tikhonov's regularization plays an important role in dealing with these problems. However, people usually pay too much attention to these problems beforehand, and hence simply apply Tikhonov's regularization to solve inverse problems. I think that the natural way to study inverse problems is to give a reconstruction procedure using the genuine observation data and examine the robustness of the reconstruction procedure when there are errors in the observation data and roundoff errors in the numerical computation. I also believe that a hybrid method, such as using a new reconstruction procedure to make a good initial guess and then using Tikhonov's regularization for the approximate identification of the unknown with high accuracy, is an effective way for solving practical inverse problems. I speculate that this method for studying inverse problems will become more popular in the near future.

Regarding my way of studying the inverse problems, we only touch on Tikhonov's regularization method a little in the next sections, but we will not discuss the existence and reconstruction using observation data with some error.

The inverse problems for elasticity are difficult, because the equations are complicated and tedious computations are required. Hence, there are only a few results for those inverse problems formulated in the previous section and their fundamental problems given in this section. Most of the known results are about the identification of an unknown boundary datum and a cavity from finitely many genuine observation data, the inverse boundary value problem having the so-called Dirichlet to Neumann map as the genuine observation data, which are obtained by infinitely many observation and the inverse scattering problem for the elastic wave. In §§5 and 6, these known results will be reviewed.

5. Identification by finite measurements

First we consider an inverse problem for identifying the unknown boundary datum. For the two-dimensional case, for example, we consider an infinitely long homogeneous isotropic elastic cylinder with uniform cross section, applied surface force and body force which are the same for each cross section and parallel to the cross section such that the cylinder is under so-called plain strain. Then, the displacement inside each cross section is the same and it satisfies the equations (1.1)-(1.3) with the range $1 \le i, j, k, l \le 2$ for indices i, j, k, l, and (1.10) given by replacing $3\lambda + 2\mu > 0$ by $\lambda + \mu > 0$. For this 2-dimensional, homogeneous, isotropic static elastic equation, we consider the inverse problem given in Section 3, Example 1. We assume that $\partial\Omega$ is smooth, B_1, B_1', B_2 are given by $B_1 u|_{\Gamma_1} = \sigma n|_{\Gamma_1}$, $B_1' u|_{\Gamma_1} = u|_{\Gamma_1}$, $B_2 u|_{\Gamma_2} = u|_{\Gamma_2}$, where σ is as given in (1.1) and n is the unit outer normal vector of $\partial\Omega$. Then, some numerical reconstruction procedures for the unknown $B_2 u|_{\Gamma_2} = g_2$ from the pair of geuine observation data ($B_1 u|_{\Gamma_1} = g_1$, $B_1' u|_{\Gamma_1} = g_1'$) were given in ([**34**], [**21**]). In [**34**], the minimization problem for the functional with regularization parameter $\alpha > 0$,

$$(5.1) \qquad J(\widetilde{g}_2) = \int_{\Gamma_1} |u(\widetilde{g}_2)(x) - g_1'(x)|^2 dx + \alpha \int_{\Omega} W(\varepsilon) dx \quad (\widetilde{g}_2 \in H^{1/2}(\Gamma_2)),$$

was solved numerically by the conjugate gradient method, where $W(\varepsilon)$ is the strain energy function for the linear strain ε defined by the displacement $u = u(\widetilde{g}_2)$, which is the solution to the mixed boundary value problem (3.3) with mixed boundary condition

$$B_1 u|_{\Gamma_1} = g_1, \quad u|_{\Gamma_2} = \widetilde{g}_2. \tag{5.2}$$

Note that the minimum solution of (5.2) is an approximate solution of g_2, called Tikhonov's regularized solution, and this method of solving the inverse problem by minimizing some functional with a regularization parameter is called the Tikhonov regularization method. In the Tikhonov regularization method, we have to solve a minimization problem for a functional, as we do here, which is not simple at all. Also, if we set the regularization parameter $\alpha = 0$, we may lose the uniqueness of the minimum solution. So how to choose $\alpha > 0$ is also a big problem.

On the other hand, in [**21**], using the iterative algorithm proposed in [**17**], an approximate solution for the unknown g_2 is numerically obtained. More precisely, for an appropriate $\widetilde{g}_2$, one first seeks the aforementioned solution $u(\widetilde{g}_2)$, then solves (3.3) with the boundary condition $B_1' u|_{\Gamma_1} = g_1'$, $\sigma(u)n|_{\Gamma_2} = \sigma(u(\widetilde{g}_2))n|_{\Gamma_2}$, and iterates this procedure. Here, $\sigma(u)$ and $\sigma(u(\widetilde{g}_2))$ are the stress tensors associated with the displacement vectors u and $u(\widetilde{g}_2)$, respectively. This iterative algorithm is a very simple algorithm that only requires one to solve the well-known forward problem iteratively, but it does not work so well for observation data with error.

Next we consider the inverse problem for identifying a cavity. That is, consider the inverse problem for identifying a cavity given in §3, Example 2, for the two-dimensional linear isotropic elastic equation. We assume that λ, $\mu \in C^3(\overline{\Omega})$, $\partial\Omega$ is smooth and B_1, B_1', B_2' are the same as in the previous inverse problem for identifying the boundary data. Then, for any $0 \neq g_1 \in C^2(\Gamma_1)$, the uniqueness for identifying γ from one set of genuine observation datum $(B_1 u|_{\Gamma_1} = g$, $B_1' u|_{\Gamma_1'} = g_1')$ holds provided that there is a solution $u \in C^3(\Omega \cup \Gamma_1 \cup \gamma) \cap C(\overline{\Omega}) \cap H^1(\Omega)$ of the associated forward problem ([**3**]).

For the inverse problem for identifying the source term, the paper [**7**] considered identifying the inelastic strain (ϵ_{kl}^*) inside a homogeneous linear isotropic static elastic medium R^3. More precisely, assuming that the density ρ is uniform and the distribution of (ϵ_{kl}^*) is confined to a bounded domain, the authors attempted to apply the Tikhonov regularization method to identify (ϵ_{kl}^*) by measuring the displacement and the traction on the surface of a ball containing the bounded domain. In engineering, for both dynamical and steady state elastic media, the inverse problem for identifying the properties of the unknown body force from the measured displacement on a part of $\partial\Omega$ or the surface inside Ω is called acoustic emission, and it is well studied ([**2**], [**33**]). These problems can be transformed to a Fredholm integral equation of the first kind by using the fundamental solution for steady state elastic media, and it is possible to apply the Tikhonov regularization method, but as far as I know there are almost no detailed mathematical studies using this approach. The method given in [**40**] can be applied to the acoustic emission problem for a dynamical elastic medium, to identify the singularity of the body force in the acoustic emission problem for a dynamical elastic medium.

Now let T be large, $\gamma = \phi$, $\Gamma_1 = \Gamma_2 = \partial\Omega$ and $B_1 = B_2$, B_1' the same as in the three-dimensional version of the inverse problem for identifying the unknown boundary datum. Assume that the density $\rho \in C^2(\overline{\Omega})$ is the only unknown for the linear isotropic dynamical elastic equation. Then, uniqueness holds for identifying the density ρ from 3 appropriately chosen pairs of genuine observation data $\{(B_1u|_{\Gamma_1} = g_1,\ B_1'u|_{\Gamma_1} = g_1')$ for 3 chosen $g_1'\}$ provided that the initial data for the associated forward problem satisfies some condition in Ω and ρ satisfies some a priori information ([**10**], [**14**]).

6. Identification by infinitely many measurements

We first introduce the Dirichlet-Neumann map Λ ((resp. $\widetilde{\Lambda}$)) as an infinite number of genuine observations. For inverse problems for the static elastic equation (3.3) ((resp. dynamical elastic equation (3.4))), let $C \in C^\infty(\overline{\Omega})$ (resp. $L_{ijkl} \in C^\infty(\overline{\Omega})$) ((resp. and also $\rho \in C^\infty(\overline{\Omega})$)), let Ω be a bounded domain with smooth boundary $\partial\Omega$, and let $\gamma = \emptyset$, $\Gamma_1 = \Gamma_2 = \partial\Omega$, b $= 0$ and $B_1u|_{\partial\Omega} = B_2u|_{\partial\Omega} = u$, $B_1'u|_{\partial\Omega} = \sigma n|_{\partial\Omega}$ ((resp. and also $\varphi_1 = \varphi_2 = 0$)), where n is the unit outer normal vector of $\partial\Omega$ and σ is given by (1.1) (resp. (2.3)). Then, define Λ ((resp. $\tilde{\Lambda}$)) by

$$\begin{aligned} &\Lambda g_1 = \sigma(u(0,g_1,0))n|_{\partial\Omega} \in H^{-1/2}(\partial\Omega) \\ &((\text{resp. } \tilde{\Lambda} g_1 = \sigma(u(0,g_1,0,0))n|_{\partial\Omega} \in L^2((0,T); H^{-1/2}(\partial\Omega))\ )) \end{aligned} \tag{6.1}$$

for any $g_1 \in H^{1/2}(\partial\Omega)$ ((resp. $g_1 \in L^2((0,T);\ H^{1/2}(\partial\Omega))$, $g_1(0) = 0$)). Here, $\sigma(u(0,g_1,0)$ ((resp. $\sigma(u(0,g_1,0,0))$)) denotes the stress tensor for the displacement vector $u(0,g_1,0)$ ((resp. $u(0,\ g_1,0,0)$)) and n is the unit outer normal vector of $\partial\Omega$.

For the linear isotropic elastic medium, λ, μ ((resp. and also ρ)) and all of their derivatives on $\partial\Omega$ can be identified uniquely from Λ ((resp. $\tilde{\Lambda}$)) ([**26**])(([**35**])). Also, not only for the linear isotropic static elastic medium but also for the linear transversally isotropic static elastic medium (([**38**])) (including the steady state case), there is a reconstruction formula for identifying C and all of their derivatives on $\partial\Omega$ from Λ ([**27**], [**29**]). There is an analogous reconstruction formula for identifying the Lamé coefficients, residual stress and all of their derivatives on $\partial\Omega$ for the isotropic static elastic medium with residual stress (including the steady state case), if we assume that the β_i $(1 \le i \le 4)$ in (2.3) are all 0 ([**30**]). Moreover, for these static elastic media, we have an approximate identification algorithm, called the layer stripping method, which approximately identifies C, the Lamé coefficients and the residual stress (for simplicity we will simply call these material coefficients from now on) ([**27**], [**29**], [**30**]). If we consider that Ω is continuously layered by thin layers, then the layer stripping method is a method which enables us to identify the material coefficients and all of their derivatives on each layer of Ω approximately step by step, starting from the shallowest layer of Ω, by stripping each layer successively; and this is why this method is called layer stripping.

The following facts are known about uniqueness for identifying the material coefficients inside Ω. For an isotropic elastic medium, we have uniqueness for identifying $\rho^{-1}\lambda$, $\rho^{-1}\mu$ from $\widetilde{\Lambda}$ ([**36**]). But for the static case, there are only partial uniqueness results for identifying the Lamé coefficients λ, μ from Λ. That is, uniqueness holds when the derivatives of μ up to a certain order are small ([**24**], [**6**], [**31**]). For simplicity, we call this assumption on μ the smallness assumption. As a byproduct of this result, an analogous uniqueness result holds for the St. Venant–Kirchhoff material ([**25**]). This is a nonlinear elastic material with a geometric nonlinearity (i.e. an elastic material which has the stress-strain relation given by (1.1) with the linear strain replaced by the strain E, where C in (1.1) is given by (1.9)), and we define Λ by using small solutions. If we consider a two-dimensional isotropic static elastic medium, uniqueness for identifying the Lamé coefficients from Λ holds if the Lamé coefficients are close to some known constant Lamé coefficient ([**23**]).

There is a general principle (not a theorem; it has to be proven for each problem) which states that if we can prove stability for identifying the unknown by using the Dirichlet-Neumann map as an observation data, then by using its proof we can replace the Dirichlet-Neumann map by some finite number of measurements (we call this replacement the discretization of the Dirichlet-Neumann map). More precisely, let V_N $(N = 1, 2, \cdots)$ be an appropriate series of finite dimensional subspaces that approximate $H^{1/2}(\partial\Omega)$, and define $\|\Lambda\|_N$ by

$$\|\Lambda\|_N := \sup_{\phi \in V_N, \|\phi\|_{H^{1/2}(\partial\Omega)} = \|\psi\|_{H^{1/2}(\partial\Omega)} = 1} |(\Lambda\phi, \psi)|.$$

Then, if Λ_j $(j = 1, 2)$ are the Dirichlet-Neumann maps for the $\lambda_j, \mu_j \in C^\infty(\overline{\Omega})$ $(j = 1, 2)$ such that their $H^s(\Omega)$ norms are bounded for some fixed large s and they satisfy the smallness assumption, then the estimate

$$\|\lambda_1 - \lambda_2\|_{L^2(\Omega)} + \|\mu_1 - \mu_2\|_{L^2(\Omega)} \leq K\eta(\|\Lambda_1 - \Lambda_2\|_{V_N} + \alpha(N))$$

is expected to hold for some constant $K > 0$. Here, $\alpha(N)$ indicates the degree of the approximation of any $g \in H^1(\partial\Omega)$ by some $g_N \in V_N$:

$$\|g - g_N\|_{H^{1/2}(\partial\Omega)} \leq \alpha(N)\|g\|_{H^1(\partial\Omega)},$$

and $\lim_{N\to\infty} \alpha(N) = 0$. Also, $\eta(\tau)$ is a function that satisfies $\lim_{\tau\to 0} \eta(\tau) = 0$. Once we have this estimate and apply the Tikhonov regularization method, it is possible to estimate the speed of convergence of Tikhonov's regularized solution to a true solution of the inverse problem in terms of the descrepancy of the error (i.e. the upper bound of the error) and the approximation index (i.e. N of V_N). Hence it is important to prove the stability estimate for identifying the Lamé coefficients from the Dirichlet-Neumann map Λ. Since it is not difficult to establish the stability estimate for identifying the Lamé coefficients on the boundary from Λ (cf. [**1**] for the two-dimensional case), we only have to prove the stability estimate for identifying the Lamé coefficients inside Ω from the Dirichlet-Neumann map. In [**5**], the author proposed an algorithm and obtained a numerical result about identifying the Lamé coefficients from the Dirichlet-Neumann map discretized in a special way. To justify this algorithm, it is expected that stability for identifying the Lamé coefficients from the Dirichlet-Neumann map is useful.

Now, let a bounded domain Ω with Lipschitz continuous boundary $\partial\Omega$ be an isotropic static elastic medium. Suppose there is an unknown inclusion D in Ω such that $\overline{D} \subset \Omega$ and $\Omega \setminus \overline{D}$ is connected. M. Ikehata showed that by using his probe method, D can be reconstructed from the localized Dirichlet-Neumann map Λ ([**13**]) provided that the Lamé coefficients λ, μ are known in $\Omega \setminus \overline{D}$ and at least one of λ, μ is discontinuous across ∂D. Also, suppose that the Lamé coefficients of D are also unknown, smooth, and satisfy the smallness assumption, and ∂D is smooth; then he showed the uniqueness of identifying the Lamé coefficients from the localized Dirichlet-Neumann map Λ. The Dirichlet-Neumann map Λ localized on an open subset $\Gamma_0 \subset \partial\Omega$ is defined by $\Lambda g_1 := \sigma(u(0, g_1, 0))n|_{\Gamma_0}$, where the support of g_1 of (6.1) is in Γ_0. The probe method is a very powerful method for showing the uniqueness and reconstruction procedure for identifying many kinds of unknown discontinuities, such as inclusions and cavities. In fact, it is possible to show the uniqueness for identifying the inclusion and reconstruction procedure for identifying the cavity in a homogeneous anisotropic elastic medium ([**11**], [**12**]).

Recalling the successful application of the probe method for identifying cracks in a conductive medium from the current to voltage map (i.e. the inverse of the Dirichlet-Neumann map), which is the idealized observation data obtained by electric impedance tomography ([**9**]), one can also use the probe method to identify cracks in an elastic medium from the Dirichlet-Neumann map.

Let Ω be the half space $\Omega = \{x = (x_1, x_2, x_3);\ x_3 > 0\}$ or the strip $\{x = (x_1, x_2, x_3);\ 0 < x_3 < H\}$. If ρ and C_{ijkl} (resp. L_{ijkl}) only depend on x_3, the elastic medium Ω is called a layered elastic medium. Usually a layered elastic medium is the term used for the case when ρ and C_{ijkl} (resp. L_{ijkl}) are piecewise constant functions of x_3. So please note that our definition of a layered elastic medium is different from the usual one. When Ω is the half space, there is a well studied inverse problem associated with the so-called Lamb problem in geophysics. This is an inverse problem for the dynamical elastic medium Ω that asks about identifying ρ, C_{ijkl} (resp. L_{ijkl}) from a finite number of input data and output data such that each of them is the traction $\delta(t)g(x_1, x_2)$ on $x_3 = 0$ and the displacement on $x_3 = 0$, respectively. By taking the Fourier transform of the displacement vector $u(t, x_1, x_2, x_3)$ with respect to (x_1, x_2), the inverse problem can be transformed to a nonlinear Volterra integral equation of the second kind with respect to the Fourier image $\hat{u}(t, \xi_1, \xi_2, x_3)$ and ρ, C_{ijkl} (resp.L_{ijkl}). For small enough $x_3 > 0$, some of ρ, C_{ijkl} (resp. L_{ijkl}) can be reconstructed by solving this integral equation ([**37**]). If we can decouple the equation and the boundary condition for $\hat{u}(t, \xi_1, \xi_2, x_3)$ so that we have 3 independent equations and boundary conditions for each 3 components of $\hat{u}(t, \xi_1, \xi_2, x_3)$, we can identify some of the unknowns in a deeper part of Ω from the surface by using the Gelfand-Levitan theory ([**18**]). The identification of residual stress near the surface for a layered isotropic elastic medium with residual stress is very important in engineering, because the residual stress is created during the manufacturing process, such as the rolling of steel to manufacture thin steel plates or hard peening to harden the surface, and the residual stress caused much trouble, such as bending after cutting a steel plate, due to the existence of the residual stress. Even for a strip, we have the same results as for a half space if we only measure for a short time and only measure on one surface (i.e. the input and output information are measured on one surface), because we do not have to consider the influence of the wave reflected from the other surface; since the residual stress is large near the surface, measuring the residual stress near the surface may be enough. In particular, we have the following results for identifying the residual stress $T(x_3)$ for the inverse problem of the Lamb problem. Namely, if the principal axis of the symmetric tensor $T(x_3)$ coincides with the coordinate axis (we will call this condition (*) from now), then we can identify most of $T(x_3)$ near the surface by measuring the displacement on the surface for a short time (i.e. the output datum) for an appropriately given $g(x_1, x_2)$ (i.e. the input datum) ([**15**]). Also, if we only restrict to identifying $T(0)$ and $\frac{d}{dx_3}T(0)$, then the assumption (*) is not necessary, and there is a reconstruction formula for identifying most of them by taking three $g(x_1, x_2)$ such that the associated three $\hat{g}(\xi_1, \xi_2)|_{\xi_1=\xi_2=0}$ are linearly independent ([**15**]).

Finally, we consider the inverse scattering problem for an elastic wave. Fixing the frequency $\omega \neq 0$, we consider the inverse scattering problem for an elastic wave in the steady state isotropic elastic medium $\Omega = R^3$. The scattering of an elastic wave here is due to the inhomogeneity of ρ, λ, μ (i.e. these are not constant), and we assume that the inhomogeneity is confined to a bounded domain in R^3. We measure the scattering amplitude $A(\theta, \theta')$ (i.e., the 3×3 matrix that determines the behavior of the scattered wave far from the bounded domain where inhomogeneity is) for the incident wave with frequency ω, incident direction θ and scattered direction θ'. Then, the inverse scattering problem is to identify $\rho^{-1}\lambda$ and $\rho^{-1}\mu$ by knowing $A(\theta, \theta')$ on ($|\theta| = |\theta'| = 1$) or a part of $|\theta| = |\theta'| = 1$. The inverse back scattering is the case when we assume that $A(\theta, \theta')$ is known on $\theta' = -\theta$, $|\theta| = 1$.

It is well known that knowing $A(\theta, \theta')$ ($|\theta| = |\theta'| = 1$) is equivalent to knowing the Dirichlet-Neumann map Λ on the boundary $\partial\Omega$ of any bounded domain Ω containing the domain where ρ, λ, μ are inhomogeneous. Hence, we have uniqueness for identifying $\rho^{-1}\lambda$, $\rho^{-1}\mu$ from the genuine observation data $A(\theta, \theta')$ ($|\theta| = |\theta'| = 1$) if we assume the smallness assumption ([**28**]). In this paper, uniqueness was proven without proving the equivalence of the genuine observation data and Λ. If $\rho = 1$ and λ, μ are very close to some constant Lamé coefficients, uniqueness for the backward inverse scattering problem is proven in ([**41**]).

The inverse scattering problem for an elastic wave scattered by a cavity can be transformed to the inverse problem for identifying the cavity from the Dirichlet-Neumann map. So this inverse problem can be solved by using the probe method, and we have the reconstruction procedure. However, the inverse scattering problem for an elastic wave by a crack is an open problem, because the inverse problem for identifying a crack from the Dirichlet-Neumann map has not yet been solved. In spite of this, there is a very interesting study in engineering ([**16**]). That is, assuming that we can parametrize the crack by a finite number of parameters, we can then compute the near field of the scattered waves (i.e. output data) for a finite number (not so many) of incident plane waves propagating from different directions (i.e. input data) by using the boundary element method, and then do the parameter estimation using some method of nonlinear programming. For the inverse scattering of an elastic wave by a crack-shaped rigid body in two-dimensional space, there is a paper by Kress ([**19**]). This work was done by using the so-called method of a complete family of surfaces (or curves), which has been successfully applied to inverse scattering problems for acoustics by Kleimann, Colton, and Kress. More precisely, by imbedding this crack-shaped rigid body in an appropriate family of curves, one sees that it is a Newton type reconstruction algorithm for the cost function with respect to the far field for the elastic wave scattered from each curve considered as a crack-shaped rigid body by computing its Fréchet derivative.

7. Summary

In this paper, I have tried to review the inverse problems for the linear elastic equation and linear isotropic elastic equation with residual stress. In sections 1 to 4, these inverse problems were formulated in general forms as much as possible, and the fundamental problems for these inverse problems were explained. In sections 5 and 6, I reviewed and explained the known and expected results for these inverse problems, and the methods that can be used to get the expected results. I am afraid that I may have explained too many things too briefly, so that readers may have difficulty reading these sections. I would like to hope that the readers can understand sections 1 to 4 at least, and just read the rest according to their interests.

I would be very happy if the readers become interested in inverse problems after reading this paper.

References

[1] M. Akamatsu, G. Nakamura and S. Steinberg, Identification of Lamé coefficients from boundary observations, Inverse Problems 7 (1991) 335-354.

[2] K. Aki and P. G. Richards, Quantitative seismology theory and mehods, vol. 1, W. H. Freeman and Company, 1980.

[3] D. D. Ang, D. D. Trong and M. Yamamoto, Identification of cavities inside two-dimensional heterogeneous isotropic elastic bodies, J. Elasticity 56 (1999) 199-212.

[4] J. Cheng and G. Nakmaura, Stability for the inverse potential problem by finite measurments on the boundary. Inverse Problems 17 (2001), 273–280.

[5] A. Constantinescu, Spectral properties and identification of elastic moduli, Inverse Problems in Engineering Mechanics, Bui, Tanaka et al. (eds), A.A. Balkema, Rotterdam (1994) 103-110.

[6] G. Eskin and J. Ralston, On the inverse boundary value problem for linear isotropic elasticity, Inverse Problems 18 (2002), 907–921.

[7] Z. Gao and T. Mura, Nonelastic strains in solids—an inverse characterization from measured boundary data, Int. J. Engrg. Sci. 30 (1992) 55-68.

[8] M. E. Gurtin, An Introduction to Continuum Mechanics, Academic Press, New York, 1981.

[9] M. Ikehata and G. Nakamura, Identification of planar crack by the probe method (preprint).

[10] M. Ikehata, G. Nakamura and M. Yamamoto, Uniqueness in inverse problems for the isotropic Lamé system, J. Math. Sci. Univ. Tokyo 5 (1998) 627-692.

[11] M. Ikehata, G. Nakamura and K. Tanuma, Identification of the shape of the inclusion in the anisotropic elastic body, Applicable Analysis 72 (1999) 17-26.

[12] M. Ikehata and G. Nakamura, Reconstruction of cavity from boundary measurements (preprint).

[13] M. Ikehata, Reconstruction of inclusion from boundary measurements, J. Inverse and Ill-Posed Problems 10 (2002), 37–65.

[14] V. Isakov, A nonhyperbolic Cauchy problem for $\square_b \square_a$ and its applications to elasticity theory, Comm. Pure Appl. Math. 39 (1986) 747-767.

[15] S. Ivanov and G. Nakamura, An algorithm for identifying the residual stress in a layered medium (preprint).

[16] S. Kobayshi, Wave analysis and boundary element method, Kyoto University Press (2000) (in Japanese).

[17] V. A. Kozlov, V. G. Maz'ya and A. F. Fomin, An iterative method for solving Cauchy problem for elliptic equations, Comput. Maths. Math. Phys., 31 (1991) 45-52.

[18] M. M. Lavrentiev, K. G. Reznitskaya and V. G. Yakhno, One-dimensional inverse problems of mathematical physics, AMS Translation Series 2, Vol. 130, AMS, Providence, RI, 1986.

[19] R. Kress, Inverse elastic scattering from a crack, Inverse Problems 12 (1996) 667-684.

[20] C-S. Man, Hartig's law and linear elasticity with initial stress, Inverse Prblems 14 (1998) 313-319.

[21] L. Marin, L. Elliott, D. B. Ingham and D. Lesnic, Boundary element method for the Cauchy problem in linear elasticity (preprint).

[22] J. E. Marsden and T. J. R. Hughes, Mathematical Foundation of Elasticity, Prentice-Hall, Inc., New Jersey, 1983.

[23] G. Nakamura and G. Uhlmann, Identification of Lamé parameters by boundary measurement, Amer. J. Math. 115 (1993) 1161-1187.

[24] G. Nakamura and G. Uhlmann, Global uniqueness for an inverse boundary problem arising in elasticity, Invent. Math. 118 (1994) 457-474.

[25] G. Nakamura and Z. Sun, An inverse boundary value problem for St. Venant-Kirchhoff materials, Inverse Problems 10 (1994) 1159-1163.

[26] G. Nakamura and G. Uhlmann, Inverse problems at the boundary for an elastic medium, SIAM J. Math. Anal. 26 (1995) 263-279.

[27] G. Nakamura and G. Uhlmann, A layer stripping algorithm in elastic impedance tomography, Inverse Problems in Wave Propagation, IMA Vol. Math. Appl. 90, Springer, New York (1997) 375-384.

[28] G. Nakamura and G. Uhlmann, Inverse scattering at a fixed energy for elasticity, J. Inv. Ill-Posed Problems, 7 (1999) 283-288.
[29] G. Nakamura, K. Tanuma and G. Uhlmann, Layer stripping for a transversely isotropic elastic medium, SIAM J. Appl. Math.59 (1999) 1879-1891.
[30] G. Nakamura and K. Tanuma, Determination of residual stress and elastic parameters from measurements at the boundary (submitted to Journal of Inverse and Ill-posed Problems).
[31] G. Nakamura and G. Uhlmann, Correction to: "Global uniqueness for an inverse boundary value problem arising in elasticity, Invent. Math. 152 (2003), 205–207.
[32] Acoustoelasticity, Japan Nondestructive Association (1994) (in Japanese).
[33] Y. Niwa and M. Ohtsu, Theory of Acoustic Emission and Its Property, Morikita (1998) (in Japanese).
[34] K. Onishi and Q. Wang, Numerical solutions of the Cauchy problem in potential and elastostatics, Inverse Problems and Related Topics, edited by G. Nakamura et al., Chapman & Hall / CRC, London (2000) 115-132.
[35] L. V. Rachele, Boundary determination for an inverse problem in elastodynamics, Comm. PDE 25 (2000), 1951–1996.
[36] L. V. Rachele, An inverse problem in elastodynamics: uniqueness of the wave speeds in the interior, J. Diff. Eq. 162 (2000) 300-325.
[37] V. G. Romanov, Inverse problems of mathematical physics, VNU Science Proess BV, 1987, Utrecht.
[38] T. C. T. Ting, Anisotropic Elasticity, Oxford Univ. Press, Oxford, 1996.
[39] N. Tosaka, K. Ohnishi and M. Yamamoto, Mathematical Method for Solving Inverse Problems, Inverse Analysis for Partial Differential Equation, Tokyo University Press (1999) (in Japanese).
[40] J-N. Wang, On the relation between singularities of coefficients and singularities of reflected waves in the Lamé system, Inverse Problems 14 (1998) 733-743.
[41] J-N. Wang, Inverse backscattering for the elastic wave equations, Comm. PDE, 25 (2000) 507-540.

DEPARTMENT OF MATHEMATICS, FACULTY OF SCIENCES, HOKKAIDO UNIVERSITY, SAPPORO 060-0810, JAPAN

Translated by GEN NAKAMURA

Amer. Math. Soc. Transl.
(2) Vol. **211**, 2003

Time-Global Solutions of Nonlinear Evolution Equations and Their Stability

Yoshihiro Shibata

1. Introduction

In this article, we consider a certain problem on stability of stationary solutions to the evolution equations. In explaining this problem, I will concentrate on my own recent results. The equation that we shall treat is

$$u_t + A(u) = f, \tag{1.1}$$

where t denotes the time variable and, as a subscript, it means partial differentiation. We denote by $x = (x_1, \cdots, x_n) \in \mathbb{R}^n$ the space variable. In general, stable states of motion mean the final equilibrium to which every perturbed solution of the evolution equation converges as the time goes to infinity. Mathematically, we first consider the following steady motion:

$$A(v) = g, \quad g = g(x), \tag{1.2}$$

where g is independent of t. We call the solution v of (1.2) the steady state solution, stationary solution or equilibrium state. In what follows, we shall discuss stability of v globally in time. More precisely, we solve (1.1) by taking $u = u(x,t) = v(x) + w(x,t)$, so that the perturbation w is governed by

$$w_t + [A(v+w) - A(v)] = f - g \tag{1.3}$$

with the initial disturbance

$$w|_{t=0} = a. \tag{1.4}$$

We say that v is *stable* if $f - g$ and a are sufficiently small that there exists a unique solution w of (1.3)–(1.4) globally in time such that $w \to 0$ as $t \to \infty$. In other worlds, we do not regard v as a stable steady state solution unless (1.3)–(1.4) possess a time-global solution for any small $f - g$ and a or unless $w \to 0$ as $t \to \infty$.

It should be noted that (1.3)–(1.4) is the initial-value problem for nonlinear evolution equations. There is now a well established perturbation theory on global solvability of nonlinear equations; our problem can be reduced to obtaining the

---

This article originally appeared in Japanese in Sūgaku **51** (1) (1999), 1–17.

2000 *Mathematics Subject Classification.* Primary 35K55; Secondary 34G20, 35B35, 47H20, 74B20, 74F05, 76D05.

decay property of solutions to the corresponding *linearized* equation. We shall explain this method, which was originally introduced in Japan by the pioneering work of Ukai [**111, 112**] and Matsumura and Nishida [**73, 74, 75, 89**]. Let us consider the linearized equation for (1.3)–(1.4):

$$w_t + A'(v(x))w = 0, \quad t > 0, \quad w|_{t=0} = b. \tag{1.5}$$

We denote by $S(t)b$ the solution of (1.5). It is important to obtain a uniform estimate of the solution to (1.5) such as

$$\|S(t)b\|_1 \leq p(t)\|b\|_2, \quad t \to \infty, \tag{1.6}$$

where $\|\cdot\|_i$ $(i = 1, 2)$ are the norms in some suitable Banach spaces. In most cases, $p(t)$ has a form like $p(t) = e^{-\gamma t}$ or $p(t) = t^{-\delta}$, where γ and δ are positive constants independent of $t > 0$. The former exhibits an exponential decay of solutions, which can be expected for the problem in *bounded* domains. The latter represents a polynomial decay of solutions, which is mainly obtained for the problem in *unbounded* domains. Once the decay property of solutions like (1.6) is established, we are ready to solve the nonlinear problem (1.3)–(1.4) as a perturbation from the linearized problem (1.5). For example, expanding (1.5) up to the second order, we obtain

$$w_t + A'(v(x))w + \int_0^1 (1-\theta)A''(v(x)+\theta w)(w,w)d\theta = f - g.$$

By the Duhamel principle, the initial-value problem can be reduced to the integral equation

$$w(t) = S(t)b - \int_0^t S(t-s)\left(\int_0^1 (1-\theta)A''(v(x)+\theta w(s))(w(s),w(s))d\theta - (f-g)\right) ds.$$

If the decay oder of $p(t)$ as $t \to \infty$ is sufficiently fast, then the contraction mapping principle enables us to obtain the global solution of w for small $f-g$ and a. Indeed, for small w, the quadratic term of w in the above integral equation becomes so small and its decay so fast that we can apply the contraction mapping principle to solve it.

To conclude this section, we mention that the stability of a steady solution cannot be obtained if the decay property of solutions to the linearized equations is unsatisfactory. Let us consider the Cauchy problem for the nonlinear wave equation

$$\begin{cases} u_{tt} = \sum_{j=1}^n \nabla \cdot \dfrac{\nabla u}{\sqrt{1+|\nabla u|^2}} & \text{for } t > 0,\ x \in \mathbb{R}^n, \\ u|_{t=0} = f(x), \quad u_t|_{t=0} = g(x), \end{cases} \tag{1.7}$$

where $\nabla = (\partial/\partial x_1, \cdots, \partial/\partial x_n)$. We discuss the stability of the trivial rest state $v \equiv 0$:

$$\sum_{j=1}^n \nabla \cdot \frac{\nabla v}{\sqrt{1+|\nabla v|^2}} = 0.$$

The linearized equation to (1.7) around $v \equiv 0$ is the following wave equation:

$$w_{tt} - \Delta w = 0,$$

where $\Delta = \sum_{j=1}^{n} \frac{\partial^2}{\partial x_j^2}$. For the solution $w = S(t)b$ of the above wave equation, von Wahl [**114**] proved that

$$\|S(t)b\|_{L^\infty} \le Ct^{-\frac{n-1}{2}}\|b\|, \quad \text{where } \|b\| \equiv \|b\|_{L^1\cap L^2} + \|\nabla b\|_{L^1\cap L^2}.$$

This implies that for $n = 1$, the solution $w(t)$ does not decay as $t \to \infty$, and that for $n = 2$, $p(t)$ in (1.6) is taken as $p(t) = t^{-\frac{1}{2}}$, which is not integrable at infinity. Indeed, for $n = 1$ and $n = 2$ Klainerman and Majda [**56**] and Hoshiga [**39**] showed, respectively, that for any small initial data f and g satisfying a certain condition, (1.7) never possesses a time-global C^2-solution. Hence it turns out that the trivial solution $v \equiv 0$ is unstable for $n = 1$ and $n = 2$. On the other hand, for the higher dimensions $n \ge 3$ it is known that $v \equiv 0$ is stable in the whole space $\mathbb{R}^n$ and in exterior domains satisfying the non-trapping condition in $\mathbb{R}^n$ ([**55**], [**57**], [**102**], [**109**]). The theory of global solvability of the Cauchy problem for the nonlinear wave equations has been remarkably developed in the last 15 years. However, investigation on stability around non-trivial solutions seems to be little established because we can expect, in general, only a slow decay estimate for solutions to the wave equation. In this article, we are mainly interested in the problem on stability around non-trivial steady solutions. From here on we will no longer mention the nonlinear wave equation.

As we have seen, the decay property of solutions to the linearized equation plays an essential role for the proof of stability of steady solutions to the nonlinear one. Therefore, it may be expected that one can deal with various kinds of nonlinear structures if we add the effect of energy dissipation to the original nonlinear wave equation. As a typical example, we add u_t to the wave equation. See Nishida [**89**], Matsumura [**73**], Nakao [**84, 85, 86**], Yamada [**115**] and the author [**103, 104**]. See also [**51, 76, 79, 80, 87, 88, 90, 92, 94**] and the references therein.

In the next section, we consider the one-dimensional *thermoelastic* equation describing the dissipative effect in thermodynamics. In recent years, the theory of the viscoelastic equation which takes Δu_t as dissipation of the viscosity has also become well established. In this direction, we refer to the excellent textbook of Racke [**98**] and the author's Prague lecture [**106**].

2. 1D thermoelastic equation

In this section, we deal with the stability problem in bounded domains. Let us consider the thermoelastic equation for the motion of elasticity with heat-conduction. The elasticity lies in the unit interval $\Omega = (0, 1)$. We denote by $X = X(t, x)$ and $T = T(t, x)$ the displacement and the temperature of the elastic bar at the point $x \in \Omega$ and the time t. The motion of the elastic bar and the energy conservation law are governed by the following equations:

$$\rho X_{tt} - S_x = f, \quad x \in \Omega, t > 0, \tag{2.1}$$

$$(e + \frac{\rho}{2}X_t^2)_t - (SX_t)_x = q_x + X_t f + g, \quad x \in \Omega,\ t > 0, \tag{2.2}$$

where S, e, q, f, g and ρ denote the force, the potential energy, the heat flux, the external force, the external supply of heat and the density, respectively. For simplicity, we assume that $\rho = 1$, $g = 0$ and $q = \kappa T_x$, where κ is a positive constant describing the thermal conductivity. As for the boundary condition, we impose either of the following ones:

$$X = x, \quad T = \tau_0 \quad \text{on } \partial\Omega = \{0\} \cup \{1\}, \tag{2.3}$$

or

$$S = 0, \quad T_x = 0 \quad \text{on } \partial\Omega. \tag{2.4}$$

The initial data is given by

$$X(0,x) = X_0(x), \quad X_t(0,x) = X_1(x), \quad T(0,x) = T_0(x), \quad 0 < x < 1. \tag{2.5}$$

Let us consider the relation between S and e. We assume that S and e are functions of X_x and T. We suppose that F is an independent variable corresponding to X_x. We denote by $\psi = \psi(F,T)$ and $\eta = \eta(F,T)$ the Helmholtz energy function and the entropy function, respectively. Suppose that S, e, ψ, η are real-valued C^∞-functions defined on $G(\delta) = \{(F,T) \in \mathbb{R}^2; |(F,T) - (1,\tau_0)| \leq \delta\}$. By the second law of thermodynamics, the following two relations are equivalent:

$$de = SdF + Td\eta \Longleftrightarrow d\psi = SdF - \eta dT, \tag{2.6}$$

which yields

$$S = \frac{d\psi}{dF}, \quad \eta = -\frac{\partial\psi}{\partial T}, \quad e = \psi - T\frac{\partial\psi}{\partial T}. \tag{2.7}$$

We impose the following hypothesis:

$$\frac{\partial^2\psi}{\partial F^2} > 0, \quad \frac{\partial^2\psi}{\partial F\partial T} \neq 0, \quad \frac{\partial^2\psi}{\partial T^2} < 0, \quad \forall(F,T) \in G(\delta). \tag{2.8}$$

Under the condition (2.7), we see that (2.2) is equivalent to

$$T\eta_t = q_x, \quad 0 < x < 1, t > 0. \tag{2.9}$$

Indeed, taking the two identities

$$\begin{aligned} e_t &= \frac{\partial\psi}{\partial T}T_t + \frac{\partial\psi}{\partial F}X_{xt} - T_t\frac{\partial\psi}{\partial T} - T\left(\frac{\partial\psi}{\partial T}\right)_t = T\eta_t + SX_{xt}, \\ \left(\frac{1}{2}X_t^2\right)_t &= X_tX_{tt} = S_xX_t + fX_t, \end{aligned}$$

into account, we obtain equivalence of (2.1)–(2.2) and (2.1)–(2.9).

Example 2.1. For $\psi(F,T) = \sqrt{1+F^2} - T^2 - \gamma TF$ with $\gamma \neq 0$, we have $S = F/\sqrt{1+F^2} - \gamma T$, $e = \sqrt{1+F^2} + T^2$, $\eta = \gamma F + 2T$. Then

$$\begin{cases} X_{tt} + (X_x/\sqrt{1+X_x^2} - \gamma T)_x = f, \\ T(2T + \gamma X_x)_t = \kappa T_{xx}. \end{cases}$$

The above example is the one-dimensional wave equation with the effect of thermal diffusion described in the Introduction. If we assume that $S(1,\tau_0) = 0$, then $X = x$, $T = \tau_0$ exhibits an equilibrium state. In general, the steady state can be described by

$$\begin{aligned} S(X_\infty'(x), T_\infty(x)) &= -F(x) + C, \quad T_\infty'' = 0, \quad 0 < x < 1, \\ F(x) &= \int_0^x f(y)dy \end{aligned} \tag{2.10}$$

with the boundary condition

$$X_\infty(x) = x, \quad T_\infty(x) = \tau_0, \quad x = 0,1, \tag{2.3'}$$

or

$$(2.4') \qquad S(X'_\infty(x), T_\infty(x)) = 0, \quad T'_\infty(x) = 0, \quad x = 0, 1.$$

In what follows, we shall discuss only the small perturbation, and hence consider the case when $C = 0$ in (2.10). For a boundary condition such as (2.4), we impose on f the condition

$$(2.11) \qquad \int_0^1 f(y)dy = 0.$$

In the case of (2.3), taking $u_\infty(x)$ as the solution of

$$S(1 + u'_\infty(x), \tau_0) = -F(x), \quad u_\infty(0) = u_\infty(1) = 0,$$

we see that $X_\infty(x) \equiv x + u_\infty(x)$, $T_\infty(x) \equiv \tau_0$ is the steady state solution. In the case of (2.4), since $T''_\infty(x) = 0$ with $T'_\infty(0) = T'_\infty(1) = 0$, we have $T_\infty(x) = T_\infty = \text{const.}$ In this case, suppose that $X_\infty(x)$ is the solution of $S(X'_\infty(x), T_\infty) = -F(x)$. Then by (2.11) we have that $S(X'_\infty(0), T_\infty) = -F(0) = 0$, $S(X'_\infty(1), T_\infty) = -F(1) = 0$, which necessarily yields the boundary condition. Furthermore, we determine $X_\infty(x)$, T_∞ from (2.2) so that the energy conservation law is fulfilled. Indeed, since $X_t f = (X_t F)_x - (X_x F)_t$, we have

$$(2.12) \qquad \int_0^1 \{e(X_x(t,x), T(t,x)) + \frac{1}{2} X_t(t,x)^2 + X_x(t,x)F(x)\}dx = e_0$$

$$\text{with} \quad e_0 \equiv \int_0^1 \{e(X'_0(x), T_0(x)) + \frac{1}{2} X_1(x)^2 + X'_0(x)F(x)\}dx.$$

Here X_0, X_1 and T_0 denote the given initial data in (2.5). Since the solution $(X(t,x), T(t,x))$ is expected to behave so that $X_t(t,x) \to 0$, $X_x(t,x) \to X'_\infty(x)$, $T(t,x) \to T_\infty$ as $t \to \infty$, we impose on $X_\infty(x)$ and T_∞ the additional condition

$$(2.13) \qquad \int_0^1 \{e(X'_\infty(x), T_\infty) + X'_\infty(x)F(x)\}dx = e_0,$$

which uniquely determines $X_\infty(x)$ and T_∞. Since (2.4) is the Neumann boundary condition, for the proof of stability we need to estimate X in terms of the first derivative of its Neumann data. Under the boundary conditions (2.3) and (2.4) we make use of the Poincaré inequality. In the latter case, we need to subtract the mean value from the right-hand side of the inequality. The mean value can be dominated by (2.12).

Let $(X_\infty(x), T_\infty)$ be a steady state solution as described above. The existence of such a solution under the conditions (2.3) and (2.4) is obtained in general for sufficiently small f. Besides smallness of f, we need to assume that, under (2.4), X'_0, X_1, T_0 are also small. Here, we state the existence theorem.

LEMMA 2.2. *(1) In the case of (2.3), there is an $\varepsilon > 0$ such that for every $f \in C^j([0,1])$ with $\max_{x\in[0,1]} |f(x)| \le \varepsilon$, there exists a unique $u_\infty \in C^{j+1}([0,1])$ satisfying*

$$S(1 + u'_\infty(x), \tau_0) = -F(x), \quad 0 < x < 1, \quad u_\infty(0) = u_\infty(1) = 0.$$

Moreover,

$$|u_\infty(x)| + |u'_\infty(x)| \le C\varepsilon, \quad 0 \le x \le 1,$$

where $C > 0$ is a constant depending only on S and τ_0.

(2) In the case of (2.4), let $(X_0(x), X_1(x), T_0(x))$ *and* f *be the initial data and the right-hand side, respectively. Suppose that* $X_0 \in H^3$, $X_1 \in H^2$, $T_0 \in H^3$, $f \in H^1$ *and that*

$$\int_0^1 f(x)dx = \int_0^1 X_0(x)dx = \int_0^1 X_1(x)dx = 0.$$

For every $\sigma > 0$ *there is an* $\varepsilon > 0$ *such that if*

$$\|(X_0', T_0) - (1, \tau_0)\|_\infty + \|X_1\|_{L^2} + \|f\|_{H^1} < \varepsilon,$$

then there exist a unique $X_\infty = X_\infty(x) \in H^3$ *and a constant* T_∞ *satisfying (2.13). Moreover,*

$$\|X_\infty' - 1\|_{H^2} + |T_\infty - \tau_0| < \sigma, \quad \text{and} \quad (X_\infty'(x), T_\infty) \in G(\delta/2), \quad 0 \le x \le 1.$$

Here

$$H^k = \{v \in L^2(0,1); dv^j/dx^j \in L^2(0,1), 0 \le j \le k\},$$
$$\|v\|_\infty = \operatorname*{ess\ sup}_{x \in (0,1)} |v(x)|.$$

We next consider the stability problem. Let $X(t,x) = X_\infty(x) + u(t,x)$, $T(t,x) = T_\infty + \theta(t,x)$. Then it follows from (2.1)–(2.5) that

(2.14)
$$\begin{cases} u_{tt} - S(X_\infty'(x) + u_x, T_\infty + \theta)_x = f(x), & 0 < x < 1, t > 0, \\ (T_\infty + \theta)\eta(X_\infty'(x) + u_x, T_\infty + \theta)_t = \kappa\theta_{xx}, & 0 < x < 1, t > 0, \\ u|_{\partial\Omega} = \theta|_{\partial\Omega} = 0 \quad \text{or} \quad S|_{\partial\Omega} = \theta_x|_{\partial\Omega} = 0, \\ u(0,x) = X_0(x) - X_\infty(x), \quad u_t(0,x) = X_1(x), \quad \theta(0,x) = T_0(x) - T_\infty. \end{cases}$$

We investigate the compatibility condition which the initial data must fulfill. Suppose that $X(t,x), T(t,x)$ is the solution. We define X_j and T_j by $X_j(x) = \partial_t^j X(0,t)$ and $T_j(x) = \partial_t^j T(0,x)$. Then X_j for $j \ge 2$ and T_j for $j \ge 1$ are successively determined by the equations starting from the initial data X_0, X_1 and T_0. For example, from (2.1) and (2.9) we obtain

$$X_2(x) = S(X_0'(x), T_0(x))_x + f(x),$$
$$T_0(x) = \left\{ \frac{\partial \eta}{\partial T}(X_0'(x), T_0(x))T_1(x) + \frac{\partial \eta}{\partial F}(X_0'(x), T_0(x))X_1'(x) \right\} = \kappa T_0''(x).$$

Here, (2.9) is expressed as

$$T\left\{ \frac{\partial \eta}{\partial T} T_t + \frac{\partial \eta}{\partial F} X_{tx} \right\} = \kappa T_{xx}.$$

Hence, according to the boundary condition as in (2.3) and (2.4), we impose the following compatibility conditions on the initial data. Under (2.3), we assume

(2.15)
$$X_j \in H^{3-j} \cap H_0^1, j = 0,1,2, \quad T_j \in H^{3-j} \cap H_0^1, j = 0,1, \quad T_2 \in L^2, \quad f \in H^1,$$

where $H_0^1 = \{v \in H^1; v|_{\partial\Omega} = 0\}$. Under (2.4), we assume

$$
\begin{aligned}
&X_j \in H^{3-j}, j = 0,1,2,3, \quad T_j \in H^{3-j}, j = 1,2, \quad T_2 \in L^2, \quad f \in H^1,\\
&S(X_0', T_0)|_{\partial\Omega} = 0, \quad \left(\frac{\partial S}{\partial F}(X_0', T_0)X_1' + \frac{\partial S}{\partial T}(X_0', T_0)T_1\right)\bigg|_{\partial\Omega} = 0,\\
&T_0'|_{\partial\Omega} = T'|_{\partial\Omega} = 0,\\
&\int_0^1 f(x)dx = \int_0^1 X_0(x)dx = \int_0^1 X_1(x)dx = 0.
\end{aligned}
\tag{2.16}
$$

These hypotheses follow necessarily from regularity of solutions together with the boundary condition, which will be stated in the next theorem.

THEOREM 2.3. *There is an* $\varepsilon > 0$ *such that if*

$$\|(X_0', T_0) - (1, \tau_0)\|_\infty + E_0 + \|f\|_{H^1} < \varepsilon,$$

then there exists a unique solution $u(t,x)$, $\theta(t,x)$ *of (2.4) with the property that*

$$u \in \bigcap_{j=0}^{3} C^j([0,\infty); H^{3-j}), \quad \theta \in \bigcap_{j=0}^{1} C^j([0,\infty); H^{3-j}), \quad \theta_{tt} \in C^0([0,\infty); L^2).$$

Moreover, such a solution is subject to a decay estimate with some absolute constants $\alpha > 0$ *and* $C > 0$*:*

$$
\begin{aligned}
&\sup_{0<s<t} e^{as}(\|\bar{D}^2(u_x, u_t, \theta(s,\cdot)\|_{L^2}^2 + \|(\theta_{xxt}, \theta_{xxx})(s,\cdot)\|_{L^2}^2)\\
&\quad + \int_0^t e^{2as}\|(D^2u, D^3u, D^1\theta, \theta_{xtt}, \theta_{xxt}, \theta_{xxx})(s,\cdot)\|_{L^2}^2 ds \le CE_0^2,
\end{aligned}
$$

where

$$E_0 = \|X_0' - X_\infty'\|_{H^2} + \|T_0 - T_\infty\|_{H^3} + \sum_{j=1}^{3} \|X_j\|_{H^{3-j}} + \|T_1\|_{H^2} + \|T_2\|_{L^2},$$

$$D^k u = \left(\frac{\partial^j u}{\partial t^k \partial x^{j-k}}, j = 0, 1, \cdots, k\right), \quad \bar{D}^k u = (u, D^1 u, \cdots, D^k u).$$

with $C(I;B)$ *denoting the* j*-times continuously differentiable functions with values in* B *on the interval* I.

The unique existence theorem on time-global solutions to the one-dimensional thermoelastic equations with the boundary condition

$$X = x, \quad T_x = 0 \quad \text{or} \quad S = 0, \quad T = \tau_0$$

had been first developed by M. Slemrod [**10**]. By means of a skillful technique of the energy method, he extended the idea of Matsumura and Nishida [**73, 74, 89**] to the thermoelastic equations. However, there remained an open problem for about eight years, since, under the boundary conditions (2.3) and (2.4), we need an essentially new breakthrough to handle complicated terms due to the boundary. To overcome such difficulty, Racke and Shibata [**100**] and Shibata [**105**] used spectral analysis to obtain a decay estimate for the linearized equations, and showed the existence of global solutions for $f = 0$ in the case of (2.3) and (2.4), respectively. The disadvantage of using spectral analysis is that it requires higher regularity for the initial data, because we need integrability of the high frequency part of the solutions. After this, Muñoz-Rivera [**82**] refined the energy method and proved an exponential decay of solutions to the linearized equations with (2.3). The key point of his idea is to estimate the boundary integrals caused by integration by parts. Indeed, as for θ, the Sobolev inequality is useful. As for u, making use of the expression of the solution, he established an estimate up to second order derivatives of the boundary integral terms, which causes a closed recurrence system. His method could be extended to nonlinear equations, and Racke-Shibata-Zheng [**101**] and Kawashima-Shibata [**53**] obtained Theorem 2.3 under the boundary conditions (2.3) and (2.4), respectively. As for the Sobolev embedding theorem, the hypothesis on H^3-regularity of the initial data is optimal to obtain C^2-solutions. In this respect, Theorem 2.3 remarkably improved the results of [**100**] and [**105**]. The same problem that Slemrod treated has been discussed by Zheng [**106**]. The Cauchy problem was investigated by Kawashima and Okada [**52**], and the problem on the half-line was developed by Jiang [**42, 43, 44, 45**] and Shibata [**108**]. The asymptotic behavior as $t \to \infty$ of the energy quantities for the solutions was obtained. In the case of a finite interval, the energy decays exponentially. For a whole or half-line, if the initial data is in $L^1 \cap L^2$, then we have

$$\|(u_x, u_t, \theta)(t, \cdot)\|_{L^1} = O(1), \quad \|(u_x, u_t, \theta)(t, \cdot)\|_{L^\infty} = O(t^{-\frac{1}{2}}),$$
$$\|(u_{xx}, u_{xt}, \theta_x, \theta_{xx})(t, \cdot)\|_{L^1} = O(t^{-\frac{1}{2}})$$

as $t \to \infty$. Kawashima [**50**] gave a deep insight into treatment of the equation from the viewpoint of the energy space.

Dafermos and Hsio [**8**], Hrusa and Messaoudi [**40**] and Hansen [**32**] showed blow-up of the solution for large initial data even though $f = 0$. Hence we cannot expect stability of the steady state solution even if it exists. Concerning the periodic solution, we refer to Feireisl [**15**] and Racke, Shibata, and Zheng [**101**]. As for decay property of solutions in terms of a different method of the energy, see Kim [**54**], Liu and Zheng [**69**], and Burns, Liu, and Zheng [**4**].

Let us consider briefly the higher dimensional case. The linearized equation of thermoelasticity has the unknown n-dimensional vector function $u = (u_1, \cdots, u_n)$ ($n = 2, 3$) and the unknown scalar function θ, which are governed by the following system:

$$\begin{cases} u_{tt} - (\lambda + u)\nabla \mathrm{div}\ u - \mu\Delta u - \gamma\nabla\theta = 0, & x \in \Omega, t > 0, \\ \theta_t - \Delta\theta - \gamma \mathrm{div}\ u_t = 0, \quad x \in \Omega, t > 0, \\ u|_{\partial\Omega} = \theta|_{\partial\Omega} = 0, \\ u(0, x) = u_0(x), \quad u_t(0, x) = u_1(x), \quad \theta(0, x) = \theta_0(x), \end{cases} \tag{2.17}$$

where $\Omega \subset \mathbb{R}^n$ is a bounded domain. Applying the operators div and rot to the first equation of (2.17), we have

$$\begin{cases} (\mathrm{div}\ u)_{tt} - (\lambda + 2\mu)\Delta(\mathrm{div}\ u) - \gamma\nabla\theta = 0, \\ \theta_t - \Delta\theta - \gamma(\mathrm{div}\ u)_t = 0, \\ (\mathrm{rot}\ u)_{tt} - u\Delta(\mathrm{rot}\ u) = 0. \end{cases}$$

This decomposition is not always useful on account of the boundary condition. For the Cauchy problem ($\Omega = \mathbb{R}^n$), however, by means of this decomposition, we obtain the energy decay and others (see Dassios and Grillakis [**12**], Muñoz-Rivera [**83**], Leis [**64**], Racke [**98**], Gawinecki [**29**]). When Ω is a bounded domain, Dafermos [**7**], Henry [**33**], Jiang, Muñoz-Rivera and Racke [**46**], and Henry, Preissinitto, and Lopes [**34**] showed that the solution of (2.17) converges to zero as $t \to \infty$ if and only if the eigenvalue problem for the vector function U,

$$-\Delta U + \lambda U = 0, \quad \text{div } U = 0, \quad x \in \Omega, \quad U|_{\partial\Omega} = 0, \tag{2.18}$$

has only the trivial solution. A sufficient condition for $U = 0$ to be the only solution of (2.18) is that the eigenvalue problem for the scalar function v,

$$-\Delta v + \lambda v = 0, \quad v|_{\partial\Omega} = 0, \tag{2.19}$$

has only simple eigenvalues. Such a situation occurs when the domain Ω has a smooth boundary $\partial\Omega$ some portion of which consists of the line. Hence in the square Ω, the solution of (2.17) tends to zero. This is not always the case when Ω is a ball. For any C^3-neighborhood of Ω, there exists a domain Ω^* in which (2.19) has only simple eigenvalues. The set of such Ω^* belongs to the second Baire category with respect to a certain C^3-distance of domains (cf. Micheletti [**77**], Uhlenbeck [**113**]). The decay rate of (u, θ) converging to zero as $t \to \infty$ is unknown in general. The rate is known for special cases such as a radially symmetric domain with radially symmetric initial data. More generally, it is known that if rot $u = 0$, then the solution decays exponentially. In such cases, the existence of time-global solutions to the corresponding *nonlinear* equations for small initial data is obtained (cf. [**46**]). During his stay in Germany in summer 1997, the author was orally informed by Prof. Racke that Lebeau and Zuazua seemed to have found a precise decay order ([**118**]). For the three-dimensional Cauchy problem of the nonlinear thermoelastic equation, Ponce and Racke [**93**] and Racke [**97**] investigated the existence and the blow-up of solutions. Time-local solutions were found by Chrzeszczyk [**6**], Jiang and Racke [**47**], Mukoyama [**81**], and Dan [**9**]. A fundamental treatment of the linearized equation can be seen in Leis [**63, 64, 65**], Racke [**95, 96**], Kupradze [**62**]. Surveys on recent topics in this direction have been published by Shibata [**106**] and Racke [**98, 99**]. In *exterior* domains in higher dimensions, there is no research on the decay property of solutions to the linearized equation and on stability or blow-up of solutions to the nonlinear equations. These problems seem to be interesting.

3. Exterior problem for the 3D Navier-Stokes equations

In this section, as an example of a stability problem in unbounded domains, we shall deal with the recent result of the author and Prof. Kobayashi (Kyushu Inst. Tech.) [**107**] on stability of solutions to the Navier-Stokes equations in an exterior domain. Suppose that the obstacle $\mathcal{O}$ moves in a 3D incompressible fluid. We take Cartesian coordinates attached to the obstacle and denote by $u = (u_1, u_2, u_3)$ and p the velocity vector and the pressure of the fluid, respectively. Then the fluid motion is governed by the following Navier-Stokes equations in the *exterior domain* $\Omega \equiv \mathbb{R}^3 \backslash \mathcal{O}$:

$$\begin{cases} u_t - \Delta u + u \cdot \nabla u + \nabla p = f, \quad \nabla \cdot u = 0, \quad x \in \Omega, t > 0, \\ u|_{\partial\Omega} = g, \quad u|_{t=0} = a, \\ \lim_{|x| \to \infty} u(t,x) = u_\infty. \end{cases} \tag{3.1}$$

For simplicity, we assume that all physical constants such as the density and the viscosity vector of the fluid are equal to one. In (3.1), u_∞ is the prescribed constant vector at infinity, and we used the notation

$$\nabla = (\frac{\partial}{\partial x_1}, \frac{\partial}{\partial x_2}, \frac{\partial}{\partial x_3}), \quad \nabla \cdot u = \text{div } u = \sum_{j=1}^{3} \frac{\partial u_j}{\partial x_j}, \quad u \cdot \nabla = \sum_{j=1}^{3} u_j \frac{\partial}{\partial x_j}.$$

The steady state problem corresponding to (3.1) is the following:

$$\begin{cases} -\Delta w + w \cdot \nabla w + \nabla q = f(x), \quad \nabla \cdot w = 0, \quad x \in \Omega, \\ w|_{\partial\Omega} = g, \\ \lim_{|x|\to\infty} = u_\infty. \end{cases} \tag{3.2}$$

In this section, we assume that f, g and u_∞ are small in a certain sense, and show the existence of the solution w and its stability. Letting $u = w + v$, $p = q + r$ in (3.1), we rewrite the original equation in terms of v and r:

(3.3)

$$\begin{cases} v_t - \Delta v + w \cdot \nabla v + v \cdot \nabla w + v \cdot \nabla v + \nabla r = 0, \quad \nabla \cdot v = 0, \quad x \in \Omega, t > 0, \\ v|_{\partial\Omega} = 0, \quad v|_{t=0} = a - w, \\ \lim_{|x|\to\infty} v(t,x) = 0. \end{cases}$$

Our problem on stability of w can now be reduced to that on the time-global existence of the solution v and r to (3.3) and on their asymptotic behavior as $t \to \infty$. Time-global existence of *weak* solutions to (3.1) was obtained by Leray [**67**] and Hopf [**38**]. However, uniqueness of weak solutions is still an open question. The existence of steady solutions w to (3.2) was shown by Leray [**66**], but it is difficult to see the asymptotic behavior at infinity of Leray's solution w. In particular, Leray's solution does not exhibit a *parabolic wake region* behind the obstacle. In this respect, Finn [**16, 18, 19, 20**] investigate the equations (3.2) and proposed the notion of *physically reasonable* (PR) solution w which behaves like

$$|w(x) - u_\infty| \le C|x|^{-\frac{1}{2}-\varepsilon} \quad \text{for some } \varepsilon > 0.$$

Finn showed the existence and uniqueness of a PR solution for small f. Finn's solution satisfies

$$|w(x) - u_\infty| \le \frac{C}{|x|} \quad \text{as } |x| \to \infty, \quad \nabla w \in L^3(\Omega).$$

He also investigated the parabolic wake region of w to some extent. Fujita [**23**] and Farwig [**13, 14**] refined Finn's work. Here we shall introduce Shibata [**107**], which deals with the estimate of the PR solution. Compared with the previous results, [**107**] improves the estimate of derivatives of the PR solution, which is necessary to show its stability.

THEOREM 3.1. *Let $3 < p < \infty$, $0 < \delta < 1/4$ and $\delta < \beta < 1 - \delta$. Suppose that $f \in L^\infty(\Omega)$, $g \in W^{2,p}(\partial\Omega)$. There is a constant $\varepsilon > 0$ depending on p, δ, β, but not on u_∞, such that if*

$$0 < |u_\infty| < \varepsilon, \quad \langle\langle f \rangle\rangle_{2\delta} + \|g\|_{W^{2,p}(\partial\Omega)} \le |u_\infty|^{\beta+\delta},$$

then (3.2) possesses a solution w, q with

$$\|w - u_\infty\|_{W^{2,p}(\Omega)} + |||w - u_\infty|||_\delta + \|q\|_{W^{1,p}(\Omega)} \le |u_\infty|^\beta. \tag{3.4}$$

Here

$$W^{k,p}(\Omega) = \{u \in L^p(\Omega); \partial_x^\alpha u \in L^p(\Omega), |\alpha| \le k\},$$

$$\langle\langle f \rangle\rangle_{2\delta} = \sup_{x\in\Omega}(1+|x|)^{\frac{5}{2}}(1+s_{u_\infty})^{\frac{1}{2}+2\delta}|f(x)|, \quad s_{u_\infty}(x) = |x| - x \cdot \frac{u_\infty}{|u_\infty|},$$

$$|||u|||_\delta = \sup_{x\in\Omega}(1+|x|)(1+s_{u_\infty}(x))^\delta|u(x)| + \sup_{x\in\Omega}(1+|x|)^{\frac{3}{2}}(1+s_{u_\infty}(x))^{\frac{1}{2}+\delta}|\nabla u(x)|.$$

In particular, it follows from (3.4) that

$$w - u_\infty \in L^{3-\delta'}(\Omega) \cap L^\infty(\Omega), \quad \nabla w \in L^{\frac{3}{2}-\delta'}(\Omega) \cap L^\infty(\Omega), \quad \text{for } 0 \le \delta' < \delta.$$

In the case when $u_\infty = 0$, Novotony and Padula [**91**], Borchers and Miyakawa [**3**], and Kozono and Yamazaki [**61**] showed existence of a solution w to (3.2) that behaves like $|w(x)| \le C/|x|$, $|\nabla w(x)| \le C/|x|^2$ as $|x| \to \infty$. This implies that $w \notin L^3(\Omega)$, $\nabla w \notin L^{3/2}(\Omega)$, but $w \in L^3_w(\Omega)$, $\nabla w \in L^{3/2}_w(\Omega)$, where L^p_w denotes the *weak* L^p-space. Hence, there is an essential difference in the behavior of solutions between the cases $u_\infty = 0$ and $u_\infty \ne 0$. The weight-function $s_{u_\infty}(x)$ exhibits a parabolic wake region behind the obstacle, which enables us to see that the decay rate as $|x| \to \infty$ of the solution $w(x)$ in the direction u_∞ is not so fast as that with $u_\infty = 0$. On the other hand, in a direction outside of the parabolic wake region, $w(x)$ decays faster than in the case of $u_\infty = 0$. Totally in the sense of integral in mean, $w(x)$ behaves much better at infinity than the solution with $u_\infty = 0$ in such a way that $w - u_\infty \in L^{3-\delta'}(\Omega) \cap L^\infty(\Omega)$, and $\nabla w \in L^{\frac{3}{2}-\delta'}(\Omega) \cap L^\infty(\Omega)$—a considerably better behavior than the case of $u_\infty = 0$.

To prove Theorem 3.1, defining $w = z + u_\infty$, we rewrite (3.2) as the following equations for z, q:

$$\text{(OS)} \qquad \begin{cases} -\Delta z + u_\infty \cdot \nabla z + z \cdot \nabla z + \nabla q = f, & \nabla \cdot z = 0 \quad \text{in } \Omega, \\ z|_{\partial\Omega} = g - u_\infty, & \lim_{|x| \to \infty} z(x) = 0. \end{cases}$$

To solve (OS), we make use of the Oseen approximation

$$-\Delta z + u_\infty \cdot \nabla z + \nabla q = f, \quad \nabla \cdot z = 0 \quad \text{in } \mathbb{R}^3.$$

The fundamental solution E of the above Oseen equation behaves at infinity like

$$|E(x)| \le \frac{C_\delta}{|x|(|u_\infty| s_{u_\infty}(x))^\delta}, \quad |\nabla E(x)| \le \frac{C_\delta}{|x|^{\frac{3}{2}} s_{u_\infty}(x)^{\frac{1}{2}} (|u_\infty| s_{u_\infty}(x))^\delta}$$

for $0 \le \delta \le 1$ with a constant C_δ independent of x. The key point of the proof of Theorem 3.1 is to show that, under some appropriate hypothesis on the right-hand side f, the volume potential $E * f$ of f defined by E exhibits the same decay rate as E as $|x| \to \infty$. To handle the boundary value problem of the Oseen equation, we may regard it as the perturbation from the Oseen approximation in $\mathbb{R}^3$ to prove that the solution is subordinate to the estimate of the solution in $\mathbb{R}^3$. Once such an estimate of the boundary value problem of the Oseen equation is established, by letting the nonlinear term $z \cdot \nabla z$ to the right-hand side of (OS), we can prove Theorem 3.1 with the aid of the contraction mapping principle. Farwig [**13, 14**] refined Finn's argument and established a list of the estimates of $E * f$ that clarifies the relations of decay at infinity between E and f. His list is very useful for research in this direction. Shibata [**107**] gave an estimate of the convolution $E * f$ by a method completely different from Farwig's [**13, 14**]. The choice between these two estimates will depend on the problem being handled. In both cases, however, the proof is so complicated that we naturally hope for a simpler one.

We shall consider the main theme, the stability of Finn's PR solutions. This stability problem was first treated by Heywood [**35**] for initial data in $L^2(\Omega)$. More precisely, he showed existence and uniqueness of a time-global solution of (3.3) for $a - w \in J^2(\Omega)$ with small $\|a - w\|_{L^2(\Omega)}$. Here $J^p(\Omega)$ is the completion in $L^p(\Omega)$ of the space $\{u \in C_0^\infty(\Omega); \nabla \cdot u = 0\}$. Furthermore, he proved that

$$\int_\Omega |\nabla v(t,x) - \nabla w(x)|^2 dx \to 0, \qquad \int_{x\in\Omega, |x|\le R} |v(t,x) - w(x)|^2 dx \to 0$$

for all $R > 0$ as $t \to \infty$. Later on, Masuda [**71**], Heywood [**36, 37**], Miyakawa [**78**], Maremonti [**70**], and Borchers and Miyakawa [**1, 2**] investigated the convergence rates. However, it is not suitable to treat the decay of solutions in the L^2-framework, which had been already pointed out by Finn [**17**]. Indeed, Finn showed that every solution w of (3.2) with $f = g = 0$ has the alternative property:

$$\text{either} \quad w \equiv 0 \quad \text{or} \quad \int_\Omega |w(x) - u_\infty|^2 dx = \infty \quad \text{for each } u_\infty \in \mathbb{R}^3.$$

Finn [**20**] proposed to solve the starting problem, which cannot be treated only by $L^2(\Omega)$. Indeed, the starting problem was solved in $L^3(\Omega)$ by Galidi, Heywood, and Shibata [**28**].

Our original problem on stability of the PR solution should be treated in the L^3-framework, which has been already proposed by Fujita and Kato [**24, 49**] and Kato [**48**]. Here we refer Kato [**48**]. We consider the case when $\Omega = \mathbb{R}^n$ $(n \ge 2)$. In this case, there is no boundary condition $u|_{\partial\Omega} = g$. Suppose that $f = 0$ and $u_\infty = 0$. Then for every $a \in J^n(\Omega)$ with sufficiently small $\|a\|_{L^n(\mathbb{R}^n)}$, there exists a unique solution $u \in C([0,\infty); L^n(\mathbb{R}^n))$ of (3.1). Moreover, such a solution has the following decay property as $t \to \infty$:

$$\text{(D)} \qquad \begin{aligned} &\|u(t,\cdot)\|_{L^p(\mathbb{R}^n)} \le C_p t^{-(\frac{1}{2}-\frac{n}{2p})}, \quad n \le p \le \infty, \\ &\|\nabla u(t,\cdot)\|_{L^n(\mathbb{R}^n)} \le C t^{-\frac{1}{2}}. \end{aligned}$$

From our point of view, this decay property implies stability of the zero solution in the L^n-framework. For exterior domains Ω in $\mathbb{R}^n$ with $n \ge 3$, Iwashita [**41**] proved (D). For the L^∞-estimate, see Chen [**5**]. If $n = 2$, then uniqueness of the Leray-Hopf solutions to (3.1) was shown by Lions and Prodi [**68**], and its decay property, such as $\lim_{t\to\infty} \|u(t)\|_{L^2(\Omega)} = 0$, was obtained by Masuda [**72**]. Later on, Kozono and Ogawa [**59, 60**] and Dan and Shibata [**10, 11**] proved (D) for $n = 2$. By these results we have stability of the rest motion with $u_\infty = 0$ in the L^n-framework in exterior domains.

However, in the case when $u_\infty \ne 0$, zero is not the solution any longer, so we need an essentially different argument. Notice that the rest motion $w \equiv 0$ is the trivial solution of (3.2) with $f = g = 0$ in the case $u_\infty = 0$. In the case $u_\infty \ne 0$, existence of a 3D steady-state solution is guaranteed by Theorem 3.1. In a 2D exterior domain, Finn and Smith [**21, 22**] and Galdi [**26, 27**] showed the existence. In the 2D case, however, the asymptotic behavior as $|x| \to \infty$ of $w(x)$ is so slow that we have more difficulty in showing its stability. Up to the present, stability of 2D flows is known only in the case of the rest motion described above. The author would like to propose considering the problem of 2D stability.

Let us now return to our original problem on the stability of Finn's PR solution constructed by Theorem 3.1. We treat this problem in the L^3-framework. Shibata [**107**] obtained the following theorem.

THEOREM 3.2. *Let $w(x)$ be the solution given by Theorem 3.1. There is an $\varepsilon > 0$ such that if $0 < |u_\infty| \le \varepsilon$ and if $a \in J^3(\Omega)$ with $\|w - a\|_{L^3(\Omega)} \le \varepsilon$, then there exists a unique time-global solution $v \in C^0([0,\infty); J^3(\Omega))$ of (3.3). Moreover, such* *a solution v has the following asymptotic behavior as $t \to \infty$:*

$$\|v(t,\cdot)\|_{L^p(\Omega)} \le C_p\sqrt{\varepsilon}t^{-(\frac{1}{2}-\frac{3}{2p})}, \quad 3 \le p \le \infty,$$
$$\|\nabla v(t,\cdot)\|_{L^3(\Omega)}\sqrt{\varepsilon}t^{-\frac{1}{2}}.$$

As is described above, in the case when $u_\infty = 0$, the steady state solution does not belong to $L^3(\Omega)$ but to $L^3_w(\Omega)$, the weak L^3-space on Ω. Hence, its stability should be discussed in $L^3_w(\Omega)$. Indeed, Borchers and Miyakawa [**3**] and Kozono and Yamazaki [**61**] showed the stability of w with $u_\infty = 0$.

Let us briefly discuss the proof of Theorem 3.2. By (3.4) in Theorem 3.1, $w - u_\infty$ is so small that we may regard (3.3) as the perturbation of the following Oseen equation:

$$\text{(3.5)} \qquad \begin{cases} u_t - \Delta u + u_\infty \cdot \nabla u + \nabla p = 0, \quad \nabla \cdot u = 0, \quad x \in \Omega, t > 0, \\ u|_{\partial\Omega} = 0, \quad u|_{t=0} = a - w, \\ \lim\limits_{|x|\to\infty} u(t,x) = 0. \end{cases}$$

One can construct the Oseen semi-group $\{T_{u_\infty}(t)\}_{t\ge 0}$ by $u(t) = T_{u_\infty}(t)(a-w)$ associated to (3.5). Miyakawa [**78**] showed that $\{T_{u_\infty}(t)\}_{t\ge 0}$ is an analytic semi-group. Moreover, Kobayashi and Shibata [**58**] established the asymptotic behavior as $t \to \infty$ for $|u_\infty| \le \sigma$:

$$\text{(3.6)} \qquad \begin{aligned} &\|T_{u_\infty}(t)a\|_{L^p(\Omega)} \le C_{\sigma,p,q}t^{-\frac{3}{2}(\frac{1}{q}-\frac{1}{p})}\|a\|_{L^q(\Omega)}, \quad 1 < q \le p \le \infty, \\ &\|\nabla T_{u_\infty}(t)a\|_{L^p(\Omega)} \le C_{\sigma,p,q}t^{-\frac{3}{2}(\frac{1}{q}-\frac{1}{p})-\frac{1}{2}}\|a\|_{L^q(\Omega)}, \quad 1 < q \le p \le 3, \end{aligned}$$

where $C_{\sigma,p,q}$ is a constant depending only on σ, p, q.

Based on (3.6), we shall prove Theorem 3.2. The proof is parallel to that of Kato [**48**]. By using the Oseen semi-group, we rewrite (3.3) as the following integral equation:

$$\text{(3.7)} \qquad v(t,\cdot) = T_{u_\infty}(t)(a-w) + L_w(t)v + N(t;v,v),$$

where

$$L_w(t)v = -\int_0^t T_{u_\infty}(t-s)P[(w-u_\infty)\cdot\nabla v(s,\cdot) + v(s,\cdot)\cdot\nabla w]ds,$$
$$N(t;v_1,v_2) = -\int_0^t T_{u_\infty}(t-s)P[v_1(s,\cdot)\cdot\nabla v_2(s,\cdot)]ds,$$
$$P : L^p(\Omega) \to J^p(\Omega); \quad \text{continuos projection.}$$

The difference of the proof between (3.7) and Kato [**48**] stems from the linear term $L_w(t)v$ on the right-hand side. For small u_∞, by using the estimate of $w - u_\infty$ in (3.4), we may regard $L_w(t)v$ as a perturbation. To this end, we need the following lemma of Poincaré-Hardy-Leray type.

LEMMA 3.3. *Let $d_\alpha(x) = |x|^{1-\alpha}s_{u_\infty}(x)^\alpha \log|x|$ for $0 \le \alpha < 3$. Then*

$$\left\|\frac{v}{d_\alpha}\right\|_{L^3(\Omega)} \le C_\alpha\|\nabla v\|_{L^3(\Omega)},$$

for all $v \in W^{1,3}(\Omega)$ with $v|_{\partial\Omega} = 0$.

By (3.4), (3.6), Lemma 3.3 and the Hölder inequality, we have

LEMMA 3.4. *Let*

$$[[v]]_{p,t} = [v]_{3,0,t} + [\nabla v]_{3,\frac{1}{2},t} + [v]_{p,\frac{1}{2}-\frac{3}{2p},t}, \quad 3 \le p < \infty,$$
$$[v]_{q,l,t} = \sup_{0 \le s \le t} s^l \|v(s,\cdot)\|_{L^q(\Omega)}.$$

Then, for every $t > 0$,

$$[[L_w(t)v]]_{p,t} \le C_{p,\delta} |||w - u_\infty|||_\delta [\nabla v]_{3,\frac{1}{2},t},$$
$$[[N(t; v_1, v_2)]]_{p,t} \le C_p [v_1]_{p,\frac{1}{2}-\frac{3}{2p},t} [\nabla v_2]_{3,\frac{1}{2},t}.$$

We take the invariant subset I defined by

$$I \equiv \{v \in C^0((0,\infty); J^3(\Omega) \cap L^p(\Omega) \cap W^{1,3}(\Omega)); [[v]]_{p,t} \le \sqrt{\varepsilon}, \quad \forall t > 0,$$
$$\text{with} \quad \lim_{t \to +0} (\|v(t,\cdot) - (a - w)\|_{L^3(\Omega)} + [\nabla v]_{3,\frac{1}{2},t} + [v]_{p,\frac{1}{2}-\frac{3}{2p},t}) = 0\}.$$

Let us consider the mapping Q defined by

$$Qv = T_{u_\infty}(t)(a - w) + L_w(t)v + N(t; v, v).$$

For sufficiently small $\varepsilon > 0$, we can show by Lemma 3.4 that Q is a contraction mapping from I into itself. The solution v of (3.7) is given by the fixed point $v \in I$ of the mapping Q. This proves Theorem 3.2.

If $f = g = 0$, then it is known that the solution given by Theorem 3.2 converges to the Kato-Iwashita solution as $u_\infty \to 0$. For the case when $(f, g) \ne 0$, this convergence is still an open question. To answer it, we need to take into account the results of Borchers and Miyakawa [**3**] and of Kozono and Yamazaki [**61**], and begin to establish the estimate of fractional powers of the Oseen operator similar to Giga [**30**], Giga and Sohr [**31**], Borchers and Miyakawa [**1, 2, 3**], and Kozono and Yamazaki [**61**]. In particular, we need to obtain a uniform estimate with respect to u_∞. This problem seems to be interesting in itself.

It is also an interesting question whether the results corresponding to Theorems 3.1 and 3.2 hold in the 2D case. However, on account of the logarithmic singularity appearing in the Stokes and Oseen resolvents, the problem is much more difficult than in the 3D case. Indeed, little is known on stability of 2D fluid motions. Proposing the question on the 2D exterior problem, we close this article.

Acknowledgement

The author would like to thank Professor Seiji Ukai (Tokyo Inst. Tech.) and Professor Mitsuhiro Nakao for their kind advice on the references. He also thanks the referee for valuable comments, and for pointing out some mistakes in the manuscript.

References

[1] W. Borchers and T. Miyakawa, *Algebraic L^2 decay for Navier–Stokes flows in exterior domains,* Acta Math. **165** (1990), 189–227.

[2] W. Borchers and T. Miyakawa, *Algebraic L^2 decay for Navier–Stokes flows in exterior domains. II,* Hiroshima Math. J. **21** (1991), 621–640.

[3] W. Borchers and T. Miyakawa, *On stability of exterior stationary Navier–Stokes flow,* Acta Math. **174** (1995), 311-382.

[4] J.A. Burns, Z. Liu and S. Zheng, *On the energy decay of a linear thermoelastic bar,* J. Math. Anal. Appl. **179** (1993), 574–591.

[5] Z.M. Chen, *Solutions of the stationary and nonstationary Navier–Stokes equations in exterior domains,* Pacific J. Math. **159** (1993), 227–240.

[6] A. Chrzeszczyk, *Some existence results in dynamical thermoelasticity, Part I. Nonlinear case,* Arch. Mech. **39** (1987), 605–617.

[7] C.M. Dafermos, *On the existence and asymptotic stability of solutions to the equations of linear thermoelasticity,* Arch. Rational Mech. Anal. **29** (1968), 241–271.

[8] C.M. Dafermos and L. Hsiao, *Development of singularities in solutions of the equations of nonlinear thermoelasticity,* Quart. Appl. Math. **44** (1986), 463–474.

[9] W. Dan, *On a local in time solvability of the Neumann problem of quasilinear hyperbolic parabolic coupled systems,* Math. Meth. Appl. Sci. **18** (1995), 1053–1082.

[10] W. Dan and Y. Shibata, *On the L_q-L_r estimates of the Stokes semigroup in a two-dimensional exterior domain,* J. Math. Soc. Japan. **51** (1999), 181–207.

[11] W. Dan and Y. Shibata, *Remark on the L_q-L_∞ estimates of the Stokes semigroup in a two-dimensional exterior domain,* Pacific J. Math. **189** (1999), 223–239.

[12] G. Dassios and M. Grillakis, *Dissipation rates and partition of energy in thermoelasticity,* Arch. Rational Mech. Anal. **87** (1984), 49–91.

[13] R. Farwig, *A variational approach in weighted Sobolev spaces to the operator $-\Delta+\partial/\partial x_1$ in exterior domains of $\mathbb{R}^3$,* Math. Z. **210** (1992), 449–464.

[14] R. Farwig, *The stationary exterior 3D-problem of Oseen and Navier–Stokes equations in anisotropically weighted Sobolev spaces,* Math. Z. **211** (1992), 409–447.

[15] E. Feireisl, *Forced vibrations in one dimensional nonlinear thermoelasticity as a local coercive-like problem,* Comment. Math. Univ. Carolinae **31** (1990), 243–255.

[16] R. Finn, *On steady-state solutions of the Navier–Stokes partial differential equations,* Arch. Rational Mech. Anal. **3** (1959), 139–151.

[17] R. Finn, *An energy theorem for viscous fluid motions,* Arch. Rational Mech. Anal. **6** (1960), 371–381.

[18] R. Finn, *On the steady-state solutions of the Navier–Stokes equations, III,* Acta Math. **105** (1961), 197–244.

[19] R. Finn, *On the exterior stationary problem for the Navier–Stokes equations and associated perturbation problems,* Arch. Rational Mech. Anal. **19** (1965), 363–406.

[20] R. Finn, *Stationary solutions of the Navier–Stokes equations,* Amer. Math. Soc., Proc. Symp. Appl. Math. **19** (1965), 121–153.

[21] R. Finn and D. R Smith, *On the linearized hydrodynamical equations in two dimensions,* Arch. Rational Mech. Anal. **25** (1967), 1–25.

[22] R. Finn and D. R Smith, *On the stationary solutions of the Navier–Stokes equations in two dimension,* Arch. Rational Mech. Anal. **25** (1967), 26–39.

[23] H. Fujita, *On the existence and regularity of the steady-state solutions of the Navier–Stokes equation,* J. Fac. Sci. Univ. Tokyo, Ser. I **9** (1961), 59–102.

[24] H. Fujita and T. Kato, *On the Navier–Stokes initial value problem I,* Arch. Rational Mech. Anal. **16** (1964), 269–315.

[25] G. P. Galdi, *Variational methods for stability of fluid motions in unbounded domains,* Ricerche Mat. **27** (1978), 387–408.

[26] G. P. Galdi, *Existence and uniqueness at low Reynolds number of stationary plane flow of a viscous fluid in exterior domains, Recent Developments in Theoretical Fluid Mechanics,* G. P. Galdi and J. Nečas eds., Pitman Research Notes in Math., vol. 291, Longman Scientific and Technical, 1993, 1–33.

[27] G. P. Galdi, *An Introduction To The Mathematical Theory Of The Navier–Stokes Equations, Vol. 1. Linearized Steady Problems; Vol. 2. Nonlinear Problems,* Springer Tracts in Natural Philosophy, vols. **38, 39** Springer-Verlag, New York, 1994.

[28] G. P. Galdi, J. G. Heywood and Y. Shibata, *On the global existence and convergence to steady state of Navier–Stokes flow past an obstacle that is started from rest,* Arch. Rational Mech. Anal. **138** (1997), 307–318.

[29] J. Gawinecki, *L^p-L^q time decay estimate for solution of the Cauchy problem for hyperbolic partial differential equations of linear thermoelasticity,* Ann. Polon. Math. **54** (1991), 135–145.

[30] Y. Giga, *Analyticity of the semigroup generated by the Stokes operator in L_r spaces,* Math. Z. **178** (1981), 297–329.

[31] Y. Giga and H. Sohr, *On the Stokes operator in exterior domains,* J. Fac. Sci. Univ. Tokyo, Sec. I A. Math. **36** (1989), 103–130.

[32] I. Hansen, *Lebensdauer von klassischen Lösungen nichtlinearer Thermolastizitätsgleichungen,* Diploma thesis, University of Bonn, 1994.
[33] D. B. Henry, *Topics in Analysis,* Publ. Secc. Mat., Univ. Autòn. Barc. **31** (1987), 29–84.
[34] D. B. Henry, A. Perissinitto Jr and O. Lopes, *On the essential spectrum of a semigroup of thermoelasticity,* Nonlinear Analysis, T. M. A. **21** (1993), 65–75.
[35] J. G. Heywood, *On stationary solutions of the Navier–Stokes equations as limits of nonstationary solutions,* Arch. Rational Mech. Anal. **37** (1970), 48–60.
[36] J. G. Heywood, *The exterior nonstationary problem for the Navier–Stokes equations,* Acta Math. **129** (1972), 11–34.
[37] J. G. Heywood, *The Navier–Stokes equations: On the existence, regularity and decay of solutions,* Indiana Univ. Math. J. **29** (1980), 639–681.
[38] E. Hopf, *Über die Anfganswertaufgabe für die hydrodynamischen Grundgleichungen,* Math. Nachr. **4** (1950–51), 213–231.
[39] A. Hoshiga, *The asymptotic behaviour of the radially symmetric solutions to quasilinear wave equations in two dimensions,* Hokkaido Math. J. **24** (1995), 575–615.
[40] W. J. Hrusa and S. A. Messaoudi, *On formation of singularities in one-dimensional nonlinear thermoelasticity,* Arch. Rational Mech. Anal. **111** (1990), 135–151.
[41] H. Iwashita, *L_q-L_r estimates for solutions of the nonstationary Stokes equations in an exterior domain and the Navier–Stokes initial value problems in L_q spaces,* Math. Ann. **285** (1989), 265–288.
[42] S. Jiang, *Global existence of smooth solutions in one-dimensional nonlinear thermoelasticity,* Proc. Roy. Soc. Edinburgh **115A** (1990), 257–274.
[43] S. Jiang, *Global solutions of the Dirichlet problem in one-dimensional nonlinear thermoelasticity,* University of Bonn, SFB 256 Preprint **138** (1990).
[44] S. Jiang, *Global solutions of Neumann problem in one-dimensional nonlinear thermoelasticity,* Nonlinear Analysis, T. M. A. **19** (1992), 107–121.
[45] S. Jiang, *On global smooth solutions to the one-dimensional equations of nonlinear inhomogeneous thermoelasticity,* Nonlinear Analysis, T. M. A. **20** (1993), 1245–1256.
[46] S. Jiang, J. E. Muñoz Rivera and R. Racke, *Asymptotic stability and global existence in thermoelasticity with symmetry,* Quart. Appl. Math. **56** (1998), 259–275.
[47] S. Jiang and R. Racke, *On some quasilinear hyperbolic-parabolic initial boundary value problems,* Math. Meth. Appl. Sci. **12** (1990), 315–339.
[48] T. Kato, *Strong L^p -solutions of the Navier–Stokes equation in $\mathbb{R}^m$ with applications to weak solutions,* Math. Z. **187** (1984), 471–480.
[49] T. Kato and H. Fujita, *On the nonstationary Navier–Stokes system,* Rend. Sem. Math. Univ. Padova **32** (1962), 243–260.
[50] S. Kawashima, *Systems of a hyperbolic-parabolic composite type, with applications to the equations of magnetohydrodynamics,* Thesis, Kyoto University, 1983.
[51] S. Kawashima, M. Nakao and K. Ono, *On the decay property of solutions to the Cauchy problem of the semilinear wave equation with a dissipative term,* J. Math. Soc. Japan **47** (1995), 617–653.
[52] S. Kawashima and M. Okada, *Smooth global solutions for the one-dimensional equations in magnetohydrodynamics,* Proc. Japan Acad. Ser. **A53** (1982), 384–387.
[53] S. Kawashima and Y. Shibata, *On the Neumann problem of one-dimensional nonlinear thermoelasticity with time-independent forces,* Czech. Math. J. **45** (1995), 39–67.
[54] J. U. Kim, *On the energy decay of a linear thermoelastic bar and plate,* SIAM J. Math. Anal. **23** (1992), 889–899.
[55] S. Klainerman, *Global existence for nonlinear wave equations,* Commun. Pure Appl. Math. **33** (1980), 43–101.
[56] S. Klainerman and A. Majda, *Formation of singularities for wave equations including the nonlinear vibrating string,* Commun. Pure Appl. Math. **33** (1980), 241–263.
[57] S. Klainerman and G. Ponce, *Global small amplitude solutions to nonlinear evolution equations,* Commun. Pure Appl. Math. **36** (1983), 133–141.
[58] T. Kobayashi and Y. Shibata, *On the Oseen equation in exterior domains,* Math. Ann. **310** (1998), 1–45.
[59] H. Kozono and T. Ogawa, *Two-dimensional Navier–Stokes flow in unbounded domains,* Math. Ann. **297** (1993), 1–31.

[60] H. Kozono and T. Ogawa, *Decay properties of strong solutions for the Navier–Stokes equations in two-dimensional unbounded domains,* Arch. Rational Mech. Anal. **122** (1993), 1–17.

[61] H. Kozono and M. Yamazaki,
(a) *On a large class of stable solutions to the Navier–Stokes equations in exterior domains,* Math. Z. **228** (1998), 751–785 ;
(b) *Exterior problem for the stationary Navier–Stokes equations in the Lorentz space,* Math. Ann. **310** (1998), 279–305.

[62] V. D. Kupradze, *Three-dimensional problems of the mathematical theory of elasticity and thermoelasticity,* North-Holland, Amsterdam, 1979.

[63] R. Leis, *Außenraumaufgaben in der linearen Elastizitätstheorie,* Math. Meth. Appl. Sci. **2** (1980), 379–396.

[64] R. Leis, *Über das asymptotische Verhalten thermoelastischer Wellen in* $\mathbb{R}^3$, Math. Meth. Appl. Sci. **3** (1981), 312–317.

[65] R. Leis, *Initial boundary value problems in mathematical physics,* John Wiley & Sons, Chichester et al., B. G. Teubner–Verlag, Stuttgart, 1986.

[66] J. Leray, *Étude de diverses équations intégrales non linéaires et de quelques problèmes que pose l'hydrodynamique,* J. Math. Pures Appl. (9) **12** (1933), 1–82.

[67] J. Leray, *Sur le mouvement d'un liquide visqueux emplissant l'espace,* Acta Math. **63** (1934), 193–248.

[68] J. L. Lions and G. Prodi, *Un théorème d'existence et unicité dans les équations de Navier–Stokes en dimension 2,* C. R. Acad. Sci. Paris **248** (1959), 3519–3521.

[69] Z. Liu and S. Zheng, *Exponential stability of the semigroup associated with a thermoelastic system,* Quart. Appl. Math. **51** (1993), 535–545.

[70] P. Maremonti, *Stabilità asintotica in media per moti fluidi viscosi in domini esterni,* Ann. Mat. Pura Appl. **97** (1985), 57–75.

[71] K. Masuda, *On the stability of incompressible viscous fluid motions past objects,* J. Math. Soc. Japan **27** (1975), 294–327.

[72] K. Masuda, *Weak solutions of Navier–Stokes equations,* Tohoku Math. **J. 36** (1984), 623–646.

[73] A. Matsumura, *Global existence and asymptotics of the solutions of the second -order quasilinear hyperbolic equations with first order dissipation,* Publ. RIMS Kyoto Univ. Series **A13** (1977), 349–379.

[74] A. Matsumura and T. Nishida, *The initial value problem for the equations of motion of compressible viscous and heat-conductive fluids,* Proc. Japan Acad. Ser. **A55** (1979), 337–342.

[75] A. Matsumura and T. Nishida, *The initial value problems for the equations of motion of viscous and heat-conductive gases,* J. Math. Kyoto Univ. **20** (1980), 67–104.

[76] T. Matsuyama and R. Ikehata, *On global solutions and energy decay for the wave equations of Kirchhoff type with nonlinear damping terms,* J. Math. Anal. Appl. **204** (1996), 729–753.

[77] A. M. Micheletti, *Perturbazione dello spettro dell'operatore di Laplace, in relazione ad una variazione del campo,* Ann. Scuola Norm. Sup. Pisa **26** (1972), 151–169.

[78] T. Miyakawa, *On nonstationary solutions of the Navier–Stokes equations in an exterior domain,* Hiroshima Math. J. **12** (1982), 115–140.

[79] K. Mochizuki and T. Motai, *On energy decay-nondecay problems for the wave equations with nonlinear dissipative term in* $\mathbb{R}^n$, J. Math. Soc. Japan **47** (1995), 405–421.

[80] M. Motai, *Asymptotic behavior of solutions to the Klein–Goldon equation with nonlinear dissipative term,* Tsukuba J. Math. **15** (1991), 151–160.

[81] T. Mukoyama, *On the Cauchy problem for quasilinear hyperbolic-parabolic coupled systems in higher dimensional spaces,* Tsukuba J. Math. **13** (1989), 363–386.

[82] J. E. Muñoz Rivera, *Energy decay rates in linear thermoelasticity,* Funkcial. Ekvac. **35** (1992), 19–30.

[83] J. E. Muñoz Rivera, *Decomposition of the displacement vector field and decay rates in linear thermoelasticity,* SIAM J. Math. Anal. **24** (1993), 1–17.

[84] M. Nakao, *Bounded, periodic or almost periodic solutions of nonlinear hyperbolic partial differential equations,* J. Diff. Eqns. **23** (1977), 368–386.

[85] M. Nakao, *Remarks on the existence and uniqueness of global decaying solutions of the nonlinear dissipative wave equations,* Math. Z. **206** (1991), 265–276.

[86] M. Nakao, *Existence of global smooth solutions to the initial-boundary value problem for the quasilinear wave equation with a degenerate dissipative term,* J. Diff. Eqns. **98** (1992), 299–327.

[87] M. Nakao and K. Ono, *Existence of global solutions to the Cauchy problem for the semilinear dissipative wave equations,* Math. Z. **214** (1993), 325–342.

[88] M. Nakao and K. Ono, *Global existence to the Cauchy problem of the semilinear wave equation with a nonlinear dissipation,* Funkcial. Ekvac. **38** (1995), 417–431.

[89] T. Nishida, *Global smooth solutions for the second order quasilinear wave equation with first order dissipation,* unpublished note (1975).

[90] K. Nishihara, *Asymptotic behavior of solutions of quasilinear hyperbolic equations with linear damping,* J. Diff. Eqns. **137** (1997), 273–301.

[91] A. Novotony and M. Padura, *Physically reasonable solutions to steady compressible Navier–Stokes equations in 3D-exterior domain ($v_\infty \neq 0$)* , Differential Integral Equations **8** (1995), 1833–1842.

[92] K. Ono, *Global existence, decay and blowup of solutions for some mildly degenerate nonlinear Kirchhoff strings,* J. Diff. Eqns. **137** (1995), 273–301.

[93] G. Ponce and R. Racke, *Global existence of small solutions to the initial value problem for nonlinear thermoelasticity,* J. Diff. Eqns. **87** (1990), 70–83.

[94] P. Pucci and J. Serrin, *Asymptotic stability for nonautonomous dissipative wave systems,* Commun. Pure Appl. Math. **69** (1996), 177–216.

[95] R. Racke, *Eigenfunction expansions in thermoelasticity,* J. Math. Anal. Appl. **120** (1986), 596–609.

[96] R. Racke, *On the time-asymptotic behavior of solutions in thermoelasticity,* Proc. Roy. Soc. Edinburgh. **107A** (1987), 289–298.

[97] R. Racke, *Blow-up in nonlinear three-dimensional thermoelasticity,* Math. Meth. Appl. Sci. **12** (1990), 267–273.

[98] R. Racke, *Lectures on nonlinear evolution equation. Initial value problems,* Aspects of Mathematics E19, Frider. Vieweg & Sohn, Braunschweig/Wiesbaden, 1992.

[99] R. Racke, *Nonlinear evolution equations in thermoelasticity,* Mathematiccal Research Note, Inst. of Maths., Univ. of Tsukuba, 96–004 June 24, 1996.

[100] R. Racke and Y. Shibata, *Global smooth solutions and asymptotic stability in one-dimensional nonlinear thermoelasticity,* Arch. Rational Mech. Anal. **116** (1991), 1–34.

[101] R. Racke, Y. Shibata and S. Zheng, *Global solvability and exponential stability in one-dimensional nonlinear thermoelasticity,* Quart. Appl. Math. **51** (1993), 751–763.

[102] J. Shatah, *Global existence of small solutions to nonlinear evolution equations,* J. Diff. Eqns. **46** (1982), 409–425.

[103] Y. Shibata, *On the global existence of classical solutions of mixed problem for some second order nonlinear hyperbolic operators with dissipative term in the interior domain,* Funkcial. Ekvac. **25** (1982), 303–345.

[104] Y. Shibata, *On the global existence of classical solutions of second order fully nonlinear hyperbolic equations with first order dissipation in the exterior domain,* Tsukuba J. Math. **7** (1983), 1–68.

[105] Y. Shibata, *Neumann problem for one-dimensional nonlinear thermoelasticity,* Partial Differential Equations (Warsaw, 1990), Banach Center Publications **27**, Polish Acad. Sci., Warsaw, 1992, 457–480.

[106] Y. Shibata, *Global in time solvability of the initial boundary value problem for some nonlinear dissipative evolution equations,* Comment. Math. Univ. Carolinae **34** (1993), 295–312.

[107] Y. Shibata, *On an exterior initial boundary value problem for Navier–Stokes equations,* Quart. Appl. Math. **57** (1999), 117–155.

[108] Y. Shibata, *On a decay rate of solutions to one dimensional thermoelastic equations on a half line; linear part,* Advances in Nonlinear Partial Differential Equations and Stochastics (S. Kawashima and T. Yanagisawa, eds.), Ser. Adv. Math. for Appl. Sci., vol. 48, World Sci. Publ., Singapore, 1998, 198–291.

[109] Y. Shibata and Y. Tsutsumi, *On a global existence theorem of small amplitude solutions for nonlinear wave equations in an exterior domain,* Math. Z. **191** (1986), 165–199.

[110] M. Slemrod, *Global existence, uniqueness, and asymptotic stability of classical smooth solutions in one-dimensional non-linear thermoelasticity,* Arch. Rational Mech. Anal. **76** (1981), 97–133.

[111] S. Ukai, *On the existence of global solutions of mixed problem for nonlinear Boltzmann equation,* Proc. Japan Acad. **50** (1974), 179–184.

[112] S. Ukai, *Les solutions globales de l'équation de Boltzmann dans l'espace tout entier et dans le demi-espace,* C. R. Acad. Paris **282** (1976), 317–320.

[113] K. Uhlenbeck, *Eigenfunctions of Laplace operators,* Bull. AMS **78** (1973), 1073–1076.

[114] W. von Wahl, *L^p-decay rates for homogeneous wave-equations,* Math. Z. **120** (1971), 93–106.

[115] Y. Yamada, *On some quasilinear wave equations with dissipative terms,* Nagoya Math. J. **187** (1982), 17–39.

[116] S. Zheng, *Global solutions and applications to a class of quasilinear hyperbolic-parabolic coupled systems,* Sci. Sinica, Ser. **A27** (1984), 1274–1286.

[117] S. Zheng and W. Shen, *Global solutions to the Cauchy problem of quasilinear hyperbolic-parabolic coupled systems,* Sci. Sinica, Ser. **A30** (1987), 1133–1149.

[118] G. Lebeau and E. Zuazua, *Decay rates for the three-dimensional system of thermoelasticity,* Arch. Rational Mech. Anal. **148** (1999), 179–231.

Department of Mathematical Sciences, Waseda University, Ohkubo, Tokyo 169-8555, Japan

Translated by HIDEO KOZONO

Amer. Math. Soc. Transl.
(2) Vol. **211**, 2003

Wavelets and Eigenvalues of Schrödinger Operators

Kazuya Tachizawa

1. Introduction

In the first half of this chapter we shall explain the basics of wavelet theory and, in the second half, their applications to the eigenvalue of the Schrödinger operators. One of the main topics in Fourier analysis is the decomposition of a function. A typical example is the Fourier series decomposition of a periodic function. In such a decomposition it is important to investigate the relation between the norm of the function and its components. For example, for the decomposition of a function in L^2 by means of a complete orthonormal system of L^2, the relation between the L^2 norm of the function and its components is given by the Parseval equality. Similarly, the relation between the L^p norm of a function in L^p and the components of its decomposition is given by the Littlewood-Paley theory. According to this theory a function in $L^p(\mathbb{R}^n)$ is decomposed into functions whose Fourier transforms are supported on $\{\xi \in \mathbb{R}^n : 2^{j-1} \leq |\xi| \leq 2^{j+1}\}$ for each $j \in \mathbb{Z}$, and the L^p norm of the function is equivalent to the L^p norm of the square function of these components (cf. [**24**]).

On the other hand, in the real-variable theory of Hardy spaces the atomic decomposition of a function was studied by Coifman and Weiss ([**9**], [**11**]). This atomic decomposition is useful in the study of Hardy spaces. For example, in order to study the boundedness of a linear operator on Hardy spaces, it is enough to investigate the behavior of the operator on atoms. After the work of Coifman and Weiss, the atomic decomposition of Littlewood-Paley type was studied by Calderón ([**4**], [**5**]) and others ([**6**], [**22**], [**60**]).

In the real analysis of Hardy spaces, Strömberg [**53**] provided an unconditional basis of Hardy spaces, which would turn out to be almost the same as the spline wavelets.

Against this background Meyer [**36**] found a complete orthonormal system in $L^2(\mathbb{R})$ consisting of Schwartz functions. After Meyer's discovery the wavelet theory

---

This article originally appeared in Japanese in Sūgaku **53** (4) (2001), 373–386.

2000 *Mathematics Subject Classification.* Primary 35P15, 42B25, 42C15, 42C40.

The author was partly supported by the Grants-in-Aid for Scientific Research, The Ministry of Education, Science, Sports and Culture, Japan.

attracted much attention and was rapidly developed ([**15**], [**16**], [**30**], [**42**], [**43**], [**44**], [**62**]).

A remarkable property of a wavelet system is that it is an unconditional basis of various function spaces. This means that a function in those function spaces can be decomposed by wavelets, and the norm of the function is characterized by the absolute values of the coefficients of the decomposition. These unconditional bases are useful in the study of function spaces. One reason a wavelet system is an unconditional basis of many spaces is that the wavelet system consists of functions that are localized in phase space. Since this localization of wavelets is suitable for the Littlewood-Paley theory, we can effectively represent a function by wavelets.

One class of operators suitable for analysis by wavelets is the Calderón-Zygmund operators. A Calderón-Zygmund operator is a generalization of the Hilbert transform or the Riesz transform. It is useful in the study of partial differential operators or pseudo-differential operators with non-smooth coefficients or symbols. An operator that is bounded on $L^2(\mathbb{R}^n)$ and has wavelets as eigenvectors and bounded eigenvalues is a Calderón-Zygmund operator. We know that a Calderón-Zygmund operator is almost diagonalized by a wavelet system. By this almost diagonalization we can prove the boundedness of the Calderón-Zygmund operator on various function spaces (cf. [**44**]).

There are many studies about the numerical analysis of partial differential operators by wavelets. In the author's opinion, wavelet theory seems to be useful in the study of partial differential operators with non-smooth coefficients. In this direction various results are known, for example, Angeletti-Mazet-Tchamitchian [**1**], Auscher-Tchamitchian [**2**], Dahlke-DeVore [**13**], Jaffard [**31**], and Tchamitchian [**57**], [**58**].

In this article an application of wavelet theory to the eigenvalue problem of Schrödinger operators is explained. The method we use originated from Fefferman's paper [**20**]. Here we restrict our concern to the Lieb-Thirring inequality, which is an estimate for the moments of negative eigenvalues of Schrödinger operators. We shall give an alternative proof of the Cwikel-Lieb-Rozenbljum inequality by means of wavelets. In our proof we shall use a wavelet characterization of weighted L^p spaces.

As wavelet theory was developed, other systems of functions whose localizations are different from wavelets have attracted attention—Wilson bases, local trigonometric bases, and Gabor frames ([**3**], [**15**], [**17**]). The function spaces that are suitable for analysis by these systems are modulation spaces of Feichtinger-Gröchenig [**21**]. Applications of these systems and modulation spaces to the theory of pseudo-differential operators are in [**28**], [**29**], [**47**], [**54**], [**55**]. In these cases it is possible to analyze the eigenvalues of the Schrödinger operators with potentials that grow to infinity at infinity.

Wavelet theory is also important in the theory of signal analysis and image analysis. In this article we do not go into these applications.

2. Wavelets

For $\psi \in L^2(\mathbb{R})$ and for $j, k \in \mathbb{Z}$ we set $\psi_{j,k}(x) = 2^{j/2}\psi(2^j x - k)$. If $\{\psi_{j,k}\}_{j,k\in\mathbb{Z}}$ is a complete orthonormal system of $L^2(\mathbb{R})$, then we say $\{\psi_{j,k}\}_{j,k\in\mathbb{Z}}$ is a wavelet basis and ψ is a wavelet ([**42**]). Before Meyer's discovery of the wavelet basis which consists of Schwartz functions in 1985, the only known wavelet basis was the Haar system, discovered in 1909. Let $\theta \in C_0^\infty(\mathbb{R})$ be an even function such that $0 \leq$

$\theta(\xi) \le 1$ for all $\xi \in \mathbb{R}$, $\operatorname{supp}\theta \subset [-4\pi/3, 4\pi/3]$, $\theta(\xi) = 1$ for all $\xi \in [-2\pi/3, 2\pi/3]$, and $\theta(\xi)^2 + \theta(2\pi - \xi)^2 = 1$ for all $\xi \in [0, 2\pi]$. Then Meyer's wavelet $\psi \in \mathcal{S}(\mathbb{R})$ is given by

$$\hat{\psi}(\xi) = \int_{\mathbb{R}} \psi(x) e^{-i\xi x}\, dx = \{\theta(\xi/2)^2 - \theta(\xi)^2\}^{1/2} e^{-i\xi/2}.$$

After Meyer's work, Lemarié [**34**] discovered the spline wavelets and Daubechies [**14**] discovered the compactly supported wavelets. These wavelets can be constructed by the framework of multiresolution analysis ([**41**]).

DEFINITION 2.1. Let $\{V_j\}_{j\in\mathbb{Z}}$ be closed subspaces in $L^2(\mathbb{R}^n)$. We say $\{V_j\}_{j\in\mathbb{Z}}$ is a *multiresolution analysis* in $L^2(\mathbb{R}^n)$ if it satisfies the following conditions:

(i) $V_j \subset V_{j+1}$ for each $j \in \mathbb{Z}$.

(ii) $\bigcap_{j\in\mathbb{Z}} V_j = \{0\}$.

(iii) $\bigcup_{j\in\mathbb{Z}} V_j$ is dense in $L^2(\mathbb{R}^n)$.

(iv) We have
$$f(x) \in V_j \iff f(2x) \in V_{j+1}$$
for each $j \in \mathbb{Z}$.

(v) There exists a $\varphi(x) \in V_0$ such that $\{\varphi(x-k)\}_{k\in\mathbb{Z}^n}$ is a complete orthonormal system of V_0.

The function φ in condition (v) is called a *scaling function.* We give some examples of multiresolution analysis in $L^2(\mathbb{R})$ (cf. [**15**], [**43**]).

EXAMPLE 2.2. Let $j \in \mathbb{Z}$ and

$$V_j = \{f \in L^2(\mathbb{R}) \,:\, f \text{ is constant on } [k2^{-j}, (k+1)2^{-j}) \text{ for each } k \in \mathbb{Z}\}.$$

Then $\{V_j\}_{j\in\mathbb{Z}}$ is a multiresolution analysis in $L^2(\mathbb{R})$. The scaling function φ is given by

$$\varphi(x) = \begin{cases} 1, & 0 \le x < 1, \\ 0, & \text{otherwise.} \end{cases}$$

EXAMPLE 2.3. Let $j \in \mathbb{Z}$ and

$$V_j = \{f \in L^2(\mathbb{R}) \cap C(\mathbb{R}) \,:\, f \text{ is linear on } [k2^{-j}, (k+1)2^{-j}) \text{ for each } k \in \mathbb{Z}\}.$$

Then $\{V_j\}_{j\in\mathbb{Z}}$ is a multiresolution analysis in $L^2(\mathbb{R})$. The scaling function φ is given by

$$\hat{\varphi}(\xi) = \left(\frac{\sin\xi/2}{\xi/2}\right)^2 \left(1 - \frac{2}{3}\sin^2\frac{\xi}{2}\right)^{-1/2}.$$

EXAMPLE 2.4. Let $\theta(\xi)$ be the function defined at the beginning of this section. Let $\varphi(x)$ be the function such that $\hat{\varphi}(\xi) = \theta(\xi)$. Then $\{\varphi(x-k)\}_{k\in\mathbb{Z}}$ is an orthonormal system in $L^2(\mathbb{R})$. Let V_0 be the L^2-closure of the linear subspace spanned by $\{\varphi(x-k)\}_{k\in\mathbb{Z}}$, and let $V_j = \{f(2^j x) \,:\, f(x) \in V_0\}$ for $j \in \mathbb{Z}$. Then $\{V_j\}_{j\in\mathbb{Z}}$ is a multiresolution analysis in $L^2(\mathbb{R})$, and φ is a scaling function.

We can construct a wavelet basis by a multiresolution analysis in $L^2(\mathbb{R})$ as follows. Let φ be the scaling function of $\{V_j\}_{j\in\mathbb{Z}}$. Let W_0 be the orthogonal complement of V_1 in V_0, and let $W_j = \{f(2^j x) \,:\, f(x) \in W_0\}$ for $j \in \mathbb{Z}$. Then W_j is the orthogonal complement of V_{j+1} in V_j, and $L^2 = \bigoplus_{j\in\mathbb{Z}} W_j$.

If we construct a $\psi \in W_0$ such that $\{\psi(x-k)\}_{k\in\mathbb{Z}}$ is a complete orthonormal system of W_0, then $\{\psi_{j,k}\}_{k\in\mathbb{Z}}$ is a complete orthonormal system of W_j. Hence $\{\psi_{j,k}\}_{j,k\in\mathbb{Z}}$ is a complete orthonormal system of $L^2(\mathbb{R})$, that is, a wavelet basis.

Since $\{2^{1/2}\varphi(2x-k)\}_{k\in\mathbb{Z}}$ is a complete orthonormal system of V_1 and $\varphi(x) \in V_0 \subset V_1$, we have

$$\varphi(x) = \sqrt{2}\sum_{k\in\mathbb{Z}} \alpha_k \varphi(2x-k);$$

that is,

$$\hat{\varphi}(\xi) = m_0\left(\frac{\xi}{2}\right)\hat{\varphi}\left(\frac{\xi}{2}\right), \qquad m_0(\xi) = \frac{1}{\sqrt{2}}\sum_{k\in\mathbb{Z}} \alpha_k e^{-ik\xi}.$$

Now we set

$$\hat{\psi}(\xi) = e^{-i\xi/2}\overline{m_0\left(\frac{\xi}{2}+\pi\right)}\hat{\varphi}\left(\frac{\xi}{2}\right);$$

that is,

$$\psi(x) = \sqrt{2}\sum_{k\in\mathbb{Z}}(-1)^{k-1}\overline{\alpha_{-k+1}}\varphi(2x-k).$$

Then ψ is a wavelet associated with $\{V_j\}_{j\in\mathbb{Z}}$.

As for the multiresolution analysis of Example 2.2, since

$$\varphi(x) = \varphi(2x) + \varphi(2x-1),$$

we have

$$\psi(x) = \varphi(2x-1) - \varphi(2x).$$

The corresponding wavelet basis $\{\psi_{j,k}\}_{j,k\in\mathbb{Z}}$ is the Haar system. The multiresolution analysis of Example 2.3 leads to the spline wavelet.

From the multiresolution analysis of Example 2.4 we get Meyer's wavelet. If we construct a multiresolution analysis with a compactly supported scaling function φ, then we get a compactly supported wavelet ([**14**]).

In general, by a multiresolution analysis in $L^2(\mathbb{R}^n)$ we can construct a corresponding wavelet basis of $L^2(\mathbb{R}^n)$ (cf. [**43**]). On the other hand, we can construct an n-dimensional wavelet basis from one-dimensional multiresolution analysis as follows. Let $\psi(x), \varphi(x)$ be a one-dimensional wavelet and its scaling function. Set $E = \{0,1\}^n \setminus \{\mathbf{0}\}$, $\psi^0(x) = \varphi(x)$, and $\psi^1(x) = \psi(x)$. For $\varepsilon = (\varepsilon_1, \ldots, \varepsilon_n) \in E$ and $x = (x_1, \ldots, x_n)$ we define

$$\psi^\varepsilon(x) = \psi^{\varepsilon_1}(x_1)\cdots\psi^{\varepsilon_n}(x_n).$$

We set $\Lambda = \{\,(\varepsilon, j, k) \,:\, \varepsilon \in E, j \in \mathbb{Z}, k \in \mathbb{Z}^n\,\}$ and

$$\psi_\lambda(x) = 2^{nj/2}\psi^\varepsilon(2^j x - k)$$

for $\lambda = (\varepsilon, j, k) \in \Lambda, x \in \mathbb{R}^n$. Then $\{\psi_\lambda\}_{\lambda\in\Lambda}$ is a complete orthonormal system of $L^2(\mathbb{R}^n)$, which is an n-dimensional wavelet basis. In the rest of this article we consider the n-dimensional wavelet basis $\{\psi_\lambda\}_{\lambda\in\Lambda}$ constructed from Meyer's wavelet.

3. Function spaces and wavelets

The wavelet basis defined in the previous section is an unconditional basis of various function spaces.

DEFINITION 3.1. Let B be a Banach space over $\mathbb{C}$. A sequence $\{e_i\}_{i=1}^{\infty} \subset B$ is an *unconditional basis* of B if it satisfies the following conditions:

(i) Any $f \in B$ has an expansion $f = \sum_{i=1}^{\infty} \alpha_i e_i$ in B, and the coefficients $\alpha_i \in \mathbb{C}$ are uniquely determined.

(ii) In the expansion of (i) we have $f = \sum_{i=1}^{\infty} \alpha_{\sigma(i)} e_{\sigma(i)}$ in B for any permutation σ of $\mathbb{N}$.

The condition (ii) is equivalent to the following condition (ii′) (see [**40**]):

(ii′) Let $f = \sum_{i=1}^{\infty} \alpha_i e_i$ be the expansion of (i). Then for all $\beta_i \in \mathbb{C}$ such that $|\beta_i| \leq |\alpha_i|$ the series $\sum_{i=1}^{\infty} \beta_i e_i$ converges in B.

The condition (ii′) suggests that if a Banach space B has an unconditional basis, then the elements of B may be characterized by the absolute values of coefficients of the element in the unconditional basis. In fact, a wavelet basis is an unconditional basis in various Banach spaces (for example, L^p spaces, Sobolev spaces, Besov spaces, Triebel-Lizorkin spaces), and an element of these spaces can be characterized by the absolute values of the wavelet coefficients of the element.

For $\lambda = (\varepsilon, j, k) \in \Lambda$ set

$$Q(\lambda) = \{(x_1, \ldots, x_n) \,:\, k_i \leq 2^j x_i < k_i + 1, \; i = 1, \ldots, n\}$$

and $\tilde{\chi}_{Q(\lambda)}(x) = |Q(\lambda)|^{-1/2} \chi_{Q(\lambda)}(x)$, where $\chi_{Q(\lambda)}(x)$ is the characteristic function of $Q(\lambda)$ and $|Q(\lambda)|$ denotes the Lebesgue measure of $Q(\lambda)$. The above cube is called a *dyadic cube*. We denote the set of all dyadic cubes in $\mathbb{R}^n$ by $\mathcal{Q}$.

First we state a theorem about L^p spaces (cf. [**43**]). We use the notation $(f, g) = \int_{\mathbb{R}^n} f \overline{g} \, dx$, and $\sim$ denotes equivalence between two norms.

THEOREM 3.2. *Let $1 < p < \infty$. Then $\{\psi_\lambda\}_{\lambda \in \Lambda}$ is an unconditional basis of $L^p(\mathbb{R}^n)$. Furthermore, for $f \in L^p(\mathbb{R}^n)$ we have $f = \sum_{\lambda \in \Lambda} (f, \psi_\lambda) \psi_\lambda$ in $L^p(\mathbb{R}^n)$ and*

$$\|f\|_{L^p} \sim \left\| \left(\sum_{\lambda \in \Lambda} (|(f, \psi_\lambda)| \tilde{\chi}_{Q(\lambda)})^2 \right)^{1/2} \right\|_{L^p}.$$

Next we define the Triebel-Lizorkin spaces and Besov spaces. Some references on these spaces are Frazier-Jawerth [**23**], Frazier-Jawerth-Weiss [**24**], Triebel [**59**] and Yamazaki [**63**]. In this article we consider homogeneous spaces.

Let ϕ be a function which satisfies the following conditions:

(i) $\phi \in \mathcal{S}(\mathbb{R}^n)$.

(ii) $\operatorname{supp} \hat{\phi} \subset \{\xi \in \mathbb{R}^n \,:\, \frac{1}{2} \leq |\xi| \leq 2\}$.

(iii) $|\hat{\phi}(\xi)| \geq c > 0$ if $\frac{3}{5} \leq |\xi| \leq \frac{5}{3}$.

For $\nu \in \mathbb{Z}$ we set $\phi_\nu(x) = 2^{\nu n}\phi(2^\nu x)$. Let $\mathcal{S}' = \mathcal{S}'(\mathbb{R}^n)$, and let $\mathcal{P}$ be the set of all polynomials on $\mathbb{R}^n$. For $\alpha \in \mathbb{R}$ and $1 \leq p, q < \infty$, we define the (homogeneous) Triebel-Lizorkin space $\dot{F}_p^{\alpha,q}$ as

$$\dot{F}_p^{\alpha,q} = \{f \in \mathcal{S}'/\mathcal{P} \ : \ \|f\|_{\dot{F}_p^{\alpha,q}} < \infty\},$$

where

$$\|f\|_{\dot{F}_p^{\alpha,q}} = \left\| \left(\sum_{\nu \in \mathbb{Z}} 2^{\alpha q \nu} |f * \phi_\nu(x)|^q \right)^{1/q} \right\|_{L^p}.$$

We also define the (homogeneous) Besov space $\dot{B}_p^{\alpha,q}$ as

$$\dot{B}_p^{\alpha,q} = \{f \in \mathcal{S}'/\mathcal{P} \ : \ \|f\|_{\dot{B}_p^{\alpha,q}} < \infty\},$$

where

$$\|f\|_{\dot{B}_p^{\alpha,q}} = \left\{ \sum_{\nu \in \mathbb{Z}} 2^{\alpha q \nu} \|f * \phi_\nu\|_{L^p}^q \right\}^{1/q}.$$

The spaces $\dot{F}_p^{\alpha,q}$ and $\dot{B}_p^{\alpha,q}$ are Banach spaces that are independent of the choice of ϕ. When $1 < p < \infty$, we can identify $\dot{F}_p^{0,2}$ with L^p. The space $\dot{F}_1^{0,2}$ is identified with the Hardy space H^1.

Let $\{\psi_\lambda\}_{\lambda \in \Lambda}$ be Meyer's wavelet basis. We remark that we can define $\langle f, \psi_\lambda \rangle$ for $f \in \mathcal{S}'/\mathcal{P}$, because the ψ_λ are Schwartz functions such that $\int_{\mathbb{R}^n} p(x)\psi_\lambda(x)\,dx = 0$ for all polynomials $p \in \mathcal{P}$.

THEOREM 3.3. *Let $\alpha \in \mathbb{R}$ and $1 \leq p, q < \infty$. Then $\{\psi_\lambda\}_{\lambda \in \Lambda}$ is an unconditional basis of $\dot{F}_p^{\alpha,q}$. For $f \in \dot{F}_p^{\alpha,q}$ we have $f = \sum_{\lambda \in \Lambda} \langle f, \psi_\lambda \rangle \psi_\lambda$ in $\dot{F}_p^{\alpha,q}$ and*

$$\|f\|_{\dot{F}_p^{\alpha,q}} \sim \left\| \left(\sum_{\lambda \in \Lambda} (|Q(\lambda)|^{-\alpha/n} |\langle f, \psi_\lambda \rangle| \tilde{\chi}_{Q(\lambda)}(x))^q \right)^{1/q} \right\|_{L^p}.$$

For each $j \in \mathbb{Z}$ we set $\Lambda_j = \{(\varepsilon, j, k) \ : \ \varepsilon \in E, k \in \mathbb{Z}^n\}$.

THEOREM 3.4. *Let $\alpha \in \mathbb{R}$ and $1 \leq p, q < \infty$. Then $\{\psi_\lambda\}_{\lambda \in \Lambda}$ is an unconditional basis of $\dot{B}_p^{\alpha,q}$. For $f \in \dot{B}_p^{\alpha,q}$ we have $f = \sum_{\lambda \in \Lambda} \langle f, \psi_\lambda \rangle \psi_\lambda$ in $\dot{B}_p^{\alpha,q}$ and*

$$\|f\|_{\dot{B}_p^{\alpha,q}} \sim \left\{ \sum_{j \in \mathbb{Z}} 2^{\alpha q j + (1/2 - 1/p) n q j} \left(\sum_{\lambda \in \Lambda_j} |\langle f, \psi_\lambda \rangle|^p \right)^{q/p} \right\}^{1/q}.$$

Similar theorems hold for inhomogeneous Triebel-Lizorkin spaces and Besov spaces (cf. [**24**], [**30**], [**43**]).

4. A_p-weights

In this section we explain A_p-weights, which will be used later. The L^p boundedness of maximal operators or singular integral operators is an important topic in harmonic analysis. The class of A_p-weights was found in the study of the boundedness of these operators on weighted L^p spaces. The properties of A_p-weights are explained in [**25**] or [**52**]. In the following a cube in $\mathbb{R}^n$ means a cube whose sides are parallel to the coordinate axes, and a weight w on $\mathbb{R}^n$ means a locally integrable function on $\mathbb{R}^n$ that is positive almost everywhere.

Let M be the Hardy-Littlewood maximal operator, that is, for $f \in L^1_{loc}(\mathbb{R}^n)$,

$$M(f)(x) = \sup_{x \in Q} \frac{1}{|Q|} \int_Q |f(y)|\, dy,$$

where the supremum is taken over all cubes Q that contain $x \in \mathbb{R}^n$.

Let w be a weight on $\mathbb{R}^n$. For $1 < p < \infty$ we say that w is an A_p-weight; that is, $w \in A_p$ if there exists a positive constant C such that

$$\frac{1}{|Q|} \int_Q w(x)\, dx \left(\frac{1}{|Q|} \int_Q w(x)^{-1/(p-1)} dx \right)^{p-1} \le C$$

for all cubes Q. For example, $w(x) = |x|^\alpha$ is an A_p-weight if $-n < \alpha < n(p-1)$.

We say that w is an A_1-weight (that is, $w \in A_1$) if there exists a positive constant C such that

$$\frac{1}{|Q|} \int_Q w\, dx \le C w(x) \quad \text{a.e. } x \in Q$$

for all cubes Q. For $1 < p < \infty$ we have $A_1 \subset A_p$. For a weight w on $\mathbb{R}^n$ we set

$$L^p(w) = \{f \,:\, f \text{ is measurable on } \mathbb{R}^n,\ \|f\|_{L^p(w)} < \infty\},$$

where

$$\|f\|_{L^p(w)} = \left\{ \int_{\mathbb{R}^n} |f(x)|^p w(x)\, dx \right\}^{1/p}.$$

In the theory of A_p-weights the following result by Muckenhoupt [**45**] is fundamental. We set $w(E) = \displaystyle\int_E w(x)\, dx$.

PROPOSITION 4.1. *Let w be a weight on $\mathbb{R}^n$.*

(i) *Let $1 < p < \infty$. Then there exists a constant $C > 0$ such that*

$$\int_{\mathbb{R}^n} M(f)(x)^p w(x)\, dx \le C \int_{\mathbb{R}^n} |f(x)|^p w(x)\, dx$$

for all $f \in L^p(w)$ if and only if $w \in A_p$.

(ii) *There exists a $C > 0$ such that*

$$w(\{x \in \mathbb{R}^n : M(f)(x) > \lambda\}) \le \frac{C}{\lambda} \int_{\mathbb{R}^n} |f(x)| w(x)\, dx$$

for all $f \in L^1(w)$ and $\lambda > 0$ if and only if $w \in A_1$.

Furthermore, the following properties hold. For a cube Q we denote by $2Q$ the cube with the same center as Q and twice the side-length.

PROPOSITION 4.2. (i) *Let $0 < \tau < 1$, and let f be a locally integrable function on $\mathbb{R}^n$ such that $M(f)(x) < \infty$ a.e. Then $(M(f))^\tau \in A_1$.*

(ii) *Let* $1 \le p < \infty$ *and* $w \in A_p$. *Then there exists a positive constant* c *such that*

$$\int_{2Q} w\,dx \le c \int_Q w\,dx$$

for all cubes $Q \subset \mathbb{R}^n$.

Property (i) is a result by Coifman and Rochberg [**10**]. As a generalization of Theorem 3.2, the following holds.

THEOREM 4.3. *Let* $1 < p < \infty$ *and* $w \in A_p$. *Then* $\{\psi_\lambda\}_{\lambda\in\Lambda}$ *is an unconditional basis of* $L^p(w)$. *Furthermore, for* $f \in L^p(w)$ *we have* $f = \sum_{\lambda\in\Lambda}(f,\psi_\lambda)\psi_\lambda$ *in* $L^p(w)$, *and*

$$\|f\|_{L^p(w)} \sim \left\| \left(\sum_{\lambda\in\Lambda} (|(f,\psi_\lambda)|\tilde{\chi}_{Q(\lambda)})^2 \right)^{1/2} \right\|_{L^p(w)}.$$

The proof of this theorem is similar to that in Lemarié-Rieusset [**35**]. For the proof we use the boundedness of Calderón-Zygmund operators on $L^p(w)$.

5. An estimate of the number of negative eigenvalues of the Schrödinger operators

In this section we explain an estimate of the moments of negative eigenvalues of Schrödinger operators, the Lieb-Thirring inequality. We shall give a proof by means of wavelets. In 1976, Lieb and Thirring proved the following theorem ([**39**]).

THEOREM 5.1. *Let* $n \ge 3$, $\gamma \ge 0$, $V \in L^{n/2+\gamma}(\mathbb{R}^n)$, *and* $V \ge 0$. *Let* $\lambda_1 \le \lambda_2 \le \cdots$ *be negative eigenvalues of the Schrödinger operator* $-\Delta - V$. *Then there exists a constant* $c > 0$, *which depends only on* n *and* γ, *such that*

$$\sum_i |\lambda_i|^\gamma \le c \int_{\mathbb{R}^n} V^{n/2+\gamma} dx.$$

The case $\gamma = 0$ in Theorem 5.1 is the following Cwikel-Lieb-Rozenbljum inequality ([**12**], [**37**], [**49**]).

THEOREM 5.2. *Let* $n \ge 3$, $V \in L^{n/2}(\mathbb{R}^n)$, *and* $V \ge 0$. *Let* N *be the number of negative eigenvalues of the Schrödinger operator* $-\Delta - V$. *Then there exists a constant* $c > 0$, *which depends only on* n, *such that*

$$N \le c \int_{\mathbb{R}^n} V^{n/2}\,dx.$$

It is known that Theorem 5.1 follows from Theorem 5.2 ([**39**]). The sharp constants of these inequalities are important in applications. See Laptev and Weidl [**33**].

By the Lieb-Thirring inequality the following Sobolev-Lieb-Thirring inequality is proved ([**18**], [**26**], [**39**]). In the statement of Theorem 5.3, $H^1(\mathbb{R}^n)$ denotes the Sobolev space of order one.

THEOREM 5.3. *Let* $n \ge 3$ *and* $n/2 < p \le 1 + n/2$. *Then there exists a positive constant* c *such that, for all orthonormal systems* $\{f_i\}_{i=1}^M$ *in* $L^2(\mathbb{R}^n)$ *such that* $\{f_i\}_{i=1}^M \subset H^1(\mathbb{R}^n)$,

$$\left\{\int_{\mathbb{R}^n}\left(\sum_{i=1}^M |f_i(x)|^2\right)^{p/(p-1)} dx\right\}^{2(p-1)/n} \leq c\sum_{i=1}^M \int_{\mathbb{R}^n} |\nabla f_i(x)|^2\,dx.$$

The Sobolev-Lieb-Thirring inequality has important applications in the problem of stability of matter in mathematical physics, and for estimates of the dimension of attractors of Navier-Stokes equations ([**38**]).

In 1983, Fefferman proved the following theorem by techniques of harmonic analysis.

THEOREM 5.4. *Let $n \geq 3$ and $r > 1$. Then there exist positive constants c_1 and c_2 such that the following holds.*

Let $V \in L^r_{loc}(\mathbb{R}^n)$, $V \geq 0$, and let $\mathcal{Q}_1$ be the set of all dyadic cubes in $\mathbb{R}^n$ such that

$$|Q|^{-2/n} \leq c_1\left(\frac{1}{|Q|}\int_Q V(x)^r\,dx\right)^{1/r}.$$

Assume that the number of minimal cubes in $\mathcal{Q}_1$ is finite. We denote the number of such minimal cubes by $\tilde{N}$. Then $N \leq c_2\tilde{N}$, where N is the number of negative eigenvalues of the Schrödinger operator $-\Delta - V$.

By this theorem we can give an alternative proof of the Cwikel-Lieb-Rozenbljum inequality. Generalizations of Theorem 5.4 and related results are in Chang-Wilson-Wolff [**7**], Chanillo-Wheeden [**8**], Kerman-Sawyer [**32**], Rochberg [**46**], Sawyer [**50**], Shen [**51**], and Wilson [**61**].

In the proof of Theorem 5.4 Fefferman used a variant of the n-dimensional Haar system. We describe it from the author's viewpoint. Let $Q_0 = \{(x_1, \ldots, x_n) : 0 \leq x_i < 1,\ i = 1, \ldots, n\}$, and let $Q_1, \ldots, Q_{2^n}$ be the half-size dyadic subcubes of Q_0. Let $E' = \{1, 2, \ldots, (n+1)(2^n - 1)\}$. Then there exist $\Psi^\varepsilon(x)$ ($\varepsilon \in E'$) that satisfy the following:

(i) For each $\varepsilon \in E'$, $\operatorname{supp}\Psi^\varepsilon \subset \overline{Q_0}$.
(ii) $\Psi^\varepsilon(x)$ is linear on each Q_i ($i = 1, \ldots, 2^n$).
(iii) For $j \in \mathbb{Z}$ and $k \in \mathbb{Z}^n$ set $\Psi^\varepsilon_{j,k}(x) = 2^{nj/2}\Psi^\varepsilon(2^j x - k)$. Then $\{\Psi^\varepsilon_{j,k} : \varepsilon \in E', j \in \mathbb{Z}, k \in \mathbb{Z}^n\}$ is a complete orthonormal system of $L^2(\mathbb{R}^n)$.

The existence of this system is proved by a method similar to the one we used when we constructed a wavelet basis by a multiresolution analysis. Fefferman's proof of Theorem 5.4 is given by weighted inequalities for $\{\Psi^\varepsilon_{j,k}\}$ similar to the equivalence of Theorem 4.3. In this article we give an alternative proof of the Cwikel-Lieb-Rozenbljum inequality by means of wavelets (cf. [**56**]).

Let $n \geq 3$, $V \in L^{n/2}(\mathbb{R}^n)$, and $V \geq 0$. Then we can realize $-\Delta - V$ as a self-adjoint operator L in $L^2(\mathbb{R}^n)$. We omit the details. We estimate the number of negative eigenvalues of L. For a number s such that $1 < s < n/2$, set $v(x) = (M(V^s)(x))^{1/s}$, where M is the Hardy-Littlewood maximal operator. We have $V(x) \leq v(x)$ a.e., and, by inequality (i) in Proposition 4.1 for $w \equiv 1$,

$$\int_{\mathbb{R}^n} v^{n/2}\,dx = \int_{\mathbb{R}^n} M(V^s)^{n/(2s)}\,dx \leq c\int_{\mathbb{R}^n} V^{n/2}\,dx < \infty.$$

Furthermore, by (i) of Proposition 4.2 we have $v \in A_1$.

For $\delta > 0$ set

$$\mathcal{I} = \{\lambda \in \Lambda : \frac{1}{|Q(\lambda)|}\int_{Q(\lambda)} v\,dx \geq \delta|Q(\lambda)|^{-2/n}\}.$$

Let $\mathcal{D}$ be the domain of L and

$$F = \{f \in \mathcal{D} : (f, \psi_\lambda) = 0 \text{ for all } \lambda \in \mathcal{I}\}.$$

By Theorem 4.3, for $f \in F$ we have

$$\int_{\mathbb{R}^n} |f|^2 V\,dx \leq \int_{\mathbb{R}^n} |f|^2 v\,dx$$
$$\leq c\sum_{\lambda \notin \mathcal{I}} |(f, \psi_\lambda)|^2 \frac{1}{|Q(\lambda)|}\int_{Q(\lambda)} v\,dx \leq c\delta \sum_{\lambda \notin \mathcal{I}} |(f, \psi_\lambda)|^2 |Q(\lambda)|^{-2/n}.$$

Now we use the following lemma.

LEMMA 5.5. *There exists a $c > 0$ such that for all $f \in H^1(\mathbb{R}^n)$*

$$\sum_{\lambda \in \Lambda} |(f, \psi_\lambda)|^2 |Q(\lambda)|^{-2/n} \leq c\int_{\mathbb{R}^n} |\nabla f(x)|^2\,dx.$$

Lemma 5.5 is proved by an argument similar to that used for the equivalence of norms in Theorem 3.3 with respect to $\dot{F}_2^{1,2}$ (cf. Meyer [**43**, p. 173]). Then, for all $f \in F$,

$$\int_{\mathbb{R}^n} |f|^2 V\,dx \leq c\delta \int_{\mathbb{R}^n} |\nabla f(x)|^2\,dx.$$

We take δ so small that $c\delta < 1$. Then for all $f \in F$ we have $(Lf, f) \geq 0$. Hence we get $N \leq \operatorname{codim} F = \sharp\mathcal{I}$, where N is the number of negative eigenvalues of L and $\sharp\mathcal{I}$ is the cardinality of $\mathcal{I}$ (cf. [**27**]). We set

$$\mathcal{B} = \{Q \in \mathcal{Q} : \frac{1}{|Q|}\int_Q v\,dx \geq \delta|Q|^{-2/n}\}.$$

Then $\sharp\mathcal{I} = (2^n - 1)\sharp\mathcal{B}$. We shall estimate $\sharp\mathcal{B}$. Let $\tilde{\mathcal{B}}$ be the set of all $Q \in \mathcal{B}$ that satisfy the following condition.

> There exists a half-size dyadic subcube $\tilde{Q}$ of Q such that all dyadic cubes contained in $\tilde{Q}$ do not belong to $\mathcal{B}$.

We have the following lemma.

LEMMA 5.6 ([**48**]). $\sharp\mathcal{B} \leq 2\sharp\tilde{\mathcal{B}}$.

Lemma 5.6 is a result about the number of vertices of the tree that is constructed by the inclusion relation between the elements of $\mathcal{B}$.

For each $Q \in \tilde{\mathcal{B}}$ we take a $\tilde{Q} \subset Q$ that satisfies the above condition. Then, by (ii) of Proposition 4.2,

$$\delta|Q|^{-2/n} \leq \frac{1}{|Q|}\int_Q v\,dx \leq \frac{1}{|Q|}\int_{3\tilde{Q}} v\,dx \leq c\frac{1}{|Q|}\int_{\tilde{Q}} v\,dx.$$

Hence by Hölder's inequality $1 \leq c\displaystyle\int_{\tilde{Q}} v^{n/2}\,dx$. Since $\{\tilde{Q}\}$ consists of mutually disjoint cubes, we get

$$N \leq (2^n - 1)\sharp\mathcal{B} \leq c\sum_{\tilde{Q}} \int_{\tilde{Q}} v^{n/2}\,dx \leq c\int_{\mathbb{R}^n} v^{n/2}\,dx \leq c\int_{\mathbb{R}^n} V^{n/2}\,dx,$$

which is the inequality in Theorem 5.1.

6. A generalization of the Lieb-Thirring inequality

In 1995, Egorov and Kondrat'ev [**19**] proved a generalization of the Lieb-Thirring inequality.

THEOREM 6.1. *Let $n \geq 3,\ \gamma \geq 0,\ q \geq n/2,\ V \geq 0$, and*

$$\int_{\mathbb{R}^n} V^{q+\gamma}|x|^{2q-n}dx < \infty.$$

Let $\lambda_1 \leq \lambda_2 \leq \cdots$ be the negative eigenvalues of the Schrödinger operator $-\Delta - V$. Then there exists a $c > 0$, which depend only on n, q, γ, such that

$$\sum_i |\lambda_i|^\gamma \leq c \int_{\mathbb{R}^n} V^{q+\gamma}|x|^{2q-n}dx.$$

Chanillo and Wheeden [**8**] proved a variant of Theorem 5.2 by replacing $-\Delta$ by second order elliptic operators that allow degenerate symbols. On the other hand, in [**19**] Egorov and Kondrat'ev proved Theorem 6.1 for higher order elliptic partial differential operators. Here we state a generalization of Theorem 6.1 for higher order elliptic partial differential operators that may have degenerate coefficients.

Let m be a natural number. We consider an operator

$$L_0 u(x) = \sum_{|\alpha|=|\beta|=m} (-1)^m D^\alpha \left(a_{\alpha\beta}(x) D^\beta u(x)\right),$$

where for $\alpha = (\alpha_1, \ldots, \alpha_n) \in (\mathbb{N} \cup \{0\})^n$

$$D^\alpha = \frac{\partial^{|\alpha|}}{\partial x_1^{\alpha_1} \cdots \partial x_n^{\alpha_n}},$$

$a_{\alpha\beta}(x) \in H^m_{loc}(\mathbb{R}^n)$, $a_{\alpha\beta} = \overline{a_{\beta\alpha}}$, and $H^m_{loc}(\mathbb{R}^n)$ is the set of all functions locally belonging to the Sobolev space $H^m(\mathbb{R}^n)$.

We assume that there exists a $w \in A_2$ such that

$$(L_0 f, f) \geq \int_{\mathbb{R}^n} w(x) \sum_{|\alpha|=m} |D^\alpha f(x)|^2\, dx$$

for all $f \in C_0^\infty(\mathbb{R}^n)$.

Then we have the following theorem ([**56**]).

THEOREM 6.2. *Let $n > 2m, q \geq n/(2m)$, and $\gamma \geq 0$. Let u be a non-negative locally integrable function on $\mathbb{R}^n$ such that $\dfrac{u}{w^q} \in A_q$. Suppose that there exists a positive constant c such that*

$$|Q|^{2m/n+1} \leq c \int_Q w\, dx \left(\int_Q \frac{u}{w^q}\, dx\right)^{1/q}$$

for all cubes $Q \subset \mathbb{R}^n$. Let V be a non-negative measurable function on $\mathbb{R}^n$ such that

$$\int_{\mathbb{R}^n} V^{q+\gamma} \frac{u}{w^q}\, dx < \infty.$$

Then the operator $L = L_0 - V$ is realized as a self-adjoint operator in $L^2(\mathbb{R}^n)$ whose negative spectrum consists only of discrete spectrum. Furthermore, there exists a positive constant c' such that

$$\sum_i |\lambda_i|^\gamma \le c' \int_{\mathbb{R}^n} V^{q+\gamma} \frac{u}{w^q}\, dx,$$

where $\{\lambda_i\}$ is the set of all negative eigenvalues of L, and c' is a constant that does not depend on V.

Theorem 6.1 is the case $L_0 = -\Delta, m = 1, w \equiv 1, u = |x|^{2q-n}$ of Theorem 6.2. In the proof of Theorem 6.2 we use a weighted version of Lemma 5.5. Theorem 6.2 is expected to be applicable to the weighted Sobolev-Lieb-Thirring inequality or estimates of the Hausdorff dimension of attractors of non-linear equations.

References

[1] J.-M. Angeletti, S. Mazet and P. Tchamitchian, *Analysis of second order elliptic operators without boundary conditions and with VMO or Hölderian coefficients*, Multiscale wavelet methods for partial differential equations, Wavelet Anal. Appl. 6, Academic Press, 1997, pp. 495–539.

[2] P. Auscher and P. Tchamitchian, *Estimates on Green's kernel using wavelets and applications*, Recent advances in wavelet analysis, Wavelet Anal. Appl. 3, Academic Press, 1994, pp. 63–86.

[3] P. Auscher, G. Weiss and M.V. Wickerhauser, *Local sine and cosine bases of Coifman and Meyer and the construction of smooth wavelets*, Wavelets—A tutorial in theory and applications (C. K. Chui, ed.), Academic Press, 1992, pp. 237–256.

[4] A.P. Calderón, *An atomic decomposition of distributions in parabolic H^p spaces*, Adv. in Math. **25** (1977), 216–225.

[5] A.P. Calderón and A. Torchinsky, *Parabolic maximal functions associated with a distribution.* I, II, Adv. in Math. **16** (1975), 1–64, and **24** (1977), 101–171.

[6] S.-Y.A. Chang and R. Fefferman, *A continuous version of duality of H^1 and BMO on the bidisc*, Ann. of Math. **112** (1980), 179–201.

[7] S.-Y.A. Chang, J.M. Wilson and T.H. Wolff, *Some weighted inequalities concerning the Schrödinger operators*, Comment. Math. Helv. **60** (1985), 217–246.

[8] S. Chanillo and R.L. Wheeden, *L^p estimates for fractional integrals and Sobolev inequalities with applications to Schrödinger operators*, Comm. Partial Differential Equation **10** (1985), 1077-1116.

[9] R. Coifman, *A real variable characterization of H^p*, Studia Math. **51** (1974), 269–274.

[10] R. Coifman and R. Rochberg, *Another characterization of BMO*, Proc. Amer. Math. Soc. **79** (1980), 249–254.

[11] R. Coifman and G. Weiss, *Extensions of Hardy spaces and their use in analysis*, Bull. Amer. Math. Soc. **83** (1977), 569–645.

[12] M. Cwikel, *Weak type estimates for singular values and the number of bound states of Schrödinger operators*, Ann. of Math. **106** (1977), 93–100.

[13] S. Dahlke and R.A. DeVore, *Besov regularity for elliptic boundary value problems*, Comm. Partial Differential Equations **22** (1997), 1–16.

[14] I. Daubechies, *Orthonormal bases of compactly supported wavelets*, Comm. Pure Appl. Math. **41** (1988), 909-996.

[15] I. Daubechies, *Ten lectures on wavelets*, SIAM, 1992.

[16] I. Daubechies ed., *Different perspectives on wavelets*, Proc. Sympos. Appl. Math. 47, Amer. Math. Soc., Providence, 1993.

[17] I. Daubechies, S. Jaffard and J-L. Journé, *A simple Wilson orthonormal basis with exponential decay*, SIAM J. Math. Anal. **22** (1991), 554-572.

[18] D.E. Edmunds and A.A. Ilyin, *On some multiplicative inequalities and approximation numbers*, Quart. J. Math. Oxford Ser.(2) **45** (1994), 159–179.

[19] Yu.V. Egorov and V.A. Kondrat'ev, *On moments of negative eigenvalues of an elliptic operator*, Math. Nachr. **174** (1995), 73–79.

[20] C. Fefferman, *The uncertainty principle*, Bull. Amer. Math. Soc. **9** (1983), 129-206.

[21] H.G. Feichtinger and K. Gröchenig, *Gabor wavelets and the Heisenberg group: Gabor expansion and short time Fourier transform from the group theoretical point of view*, Wavelets—A tutorial in theory and applications (C. K. Chui, ed.), Academic Press, 1992, pp. 359–397.

[22] M. Frazier and B. Jawerth, *Decomposition of Besov spaces*, Indiana Univ. Math. J. **34** (1985), 777–799.

[23] M. Frazier and B. Jawerth, *A discrete transform and decompositions of distribution spaces*, J. Funct. Anal. **93** (1990), 34-170.

[24] M. Frazier, B. Jawerth and G. Weiss, *Littlewood-Paley theory and the study of function spaces*, CBMS Regional Conference Series in Mathematics 79, Amer. Math. Soc., 1991.

[25] J. Garcia-Cuerva and J.L. Rubio de Francia, *Weighted norm inequalities and related topics*, North-Holland Math. Stud. 116, North-Holland, Amsterdam, 1985.

[26] J.-M. Ghidaglia, M. Marion, and R. Temam, *Generalization of the Sobolev-Lieb-Thirring inequalities and applications to the dimension of attractors*, Differential Integral Equations **1** (1988), 1–21.

[27] I.M. Glazman, *Direct methods of qualitative spectral analysis of singular differential operators*, Daniel Davey & Co., Inc., 1965.

[28] K. Gröchenig and C. Heil, *Modulation spaces and pseudodifferential operators*, Integral Equations Operator Theory **34** (1999), 439–457.

[29] C. Heil, J. Ramanathan and P. Topiwala, *Singular values of compact pseudodifferential operators*, J. Funct. Anal. **150** (1997), 426–452.

[30] E. Hernández and G. Weiss, *A first course on wavelets*, CRC Press, 1996.

[31] S. Jaffard, *Analyse par ondelettes d'un problème elliptique singulier*, J. Math. Pures Appl. **72** (1993), 121–143.

[32] R. Kerman and E. Sawyer, *The trace inequality and eigenvalue estimates for Schrödinger operators*, Ann. Inst. Fourier (Grenoble) **36** (1986), 207–228.

[33] A. Laptev and T. Weidl, *Sharp Lieb-Thirring inequalities in higher dimensions*, Acta Math. **184** (2000), 87–111.

[34] P.G. Lemarié, *Ondelettes à localisation exponentielle*, J. Math. Pures Appl. **67** (1988), 227–236.

[35] P.G. Lemarié-Rieusset, *Ondelettes et poids de Muckenhoupt*, Studia Math. **108** (1994), 127–147.

[36] P.G. Lemarié and Y. Meyer, *Ondelettes et bases hilbertiennes*, Rev. Mat. Iberoamericana **2** (1986), 1–18.

[37] E. Lieb, *Bounds on the number of eigenvalues of the Laplace and Schrödinger operators*, Bull. Amer. Math. Soc. **82** (1976), 751-753.

[38] E. Lieb, *Bounds on Schrödinger operators and generalized Sobolev-type inequalities with applications in mathematics and physics*, Inequalities, Lecture Notes in Pure and Appl. Math. 129, Dekker, 1991, pp. 123–133.

[39] E. Lieb and W. Thirring, *Inequalities for the moments of the eigenvalues of the Schrödinger Hamiltonian and their relation to Sobolev inequalities*, Studies in mathematical physics, Essays in honor of V. Bargmann, Princeton University Press, 1976, pp. 269–303.

[40] J. Lindenstrauss and L. Tzafriri, *Classical Banach spaces*. I, Springer-Verlag, 1977.

[41] S. Mallat, *Multiresolution approximations and wavelet orthonormal bases of* $L^2(\mathbb{R}^n)$, Trans. Amer. Math. Soc. **315** (1989), 69–87.

[42] Y. Meyer, *Wavelets : algorithms and applications*, SIAM, 1993.

[43] Y. Meyer, *Wavelets and operators*, Cambridge Studies in Advanced Mathematics 37, Cambridge University Press, Cambridge, 1992.

[44] Y. Meyer and R. Coifman, *Wavelets. Calderón-Zygmund and multilinear operators*, Cambridge Studies in Advanced Mathematics 48, Cambridge University Press, Cambridge, 1997.

[45] B. Muckenhoupt, *Weighted norm inequalities for the Hardy maximal function*, Trans. Amer. Math. Soc. **165** (1972), 207–226.

[46] R. Rochberg, *NWO sequences, weighted potential operators, and Schrödinger eigenvalues*, Duke Math. J. **72** (1993), 187–215.

[47] R. Rochberg and K. Tachizawa, *Pseudodifferential operators, Gabor frames, and local trigonometric bases*, H. Feichtinger and T. Strohmer, eds. Gabor analysis and algorithms: Theory and applications, Birkhäuser, 1998, pp. 171–192.

[48] R. Rochberg and M. Taibleson, *An averaging operator on a tree*, Harmonic analysis and partial differential equations (El Escorial, 1987), Springer Lecture Notes, 1384, pp. 207-213.

[49] G.V. Rozenbljum, *Distribution of the discrete spectrum of singular differential operators*, Soviet Math. Dokl. **13** (1972), 245–249.

[50] E. Sawyer, *A weighted inequality and eigenvalue estimates for Schrödinger operators*, Indiana Univ. Math. J. **35** (1986), 1–28.

[51] Z. Shen, *On the number of negative eigenvalues for a Schrödinger operator with magnetic field*, Comm. Math. Phys. **182** (1996), 637–660.

[52] E.M. Stein, *Harmonic analysis: Real-variable methods, orthogonality, and oscillatory integrals*, Princeton Mathematical Series, 43, Princeton University Press, 1993.

[53] J.-O. Strömberg, *A modified Franklin system and higher-order spline systems on $\mathbb{R}^n$ as unconditional bases for Hardy spaces*, Conf. in Harmonic Analysis in Honor of Antoni Zygmund, vol II (Beckner et al., eds.), Univ. of Chicago Press, 1981, pp. 475–494.

[54] K. Tachizawa, *The boundedness of pseudodifferential operators on modulation spaces*, Math. Nachr. **168** (1994), 263–277.

[55] K. Tachizawa, *The pseudodifferential operators and Wilson bases*, J. Math. Pures et Appl. **75** (1996), 509–529.

[56] K. Tachizawa, *On the moments of the negative eigenvalues of elliptic operators*, J. Fourier Anal. Appl. **8** (2002), 233–244.

[57] P. Tchamitchian, *Wavelets and differential operators*, Different perspectives on wavelets, Proc. Sympos. Appl. Math. 47, Amer. Math. Soc., Providence, 1993, pp. 77–88.

[58] P. Tchamitchian, *Inversion de certains operateurs elliptiques à coefficients variables*, SIAM J. Math. Anal. **27** (1996), 1680–1703.

[59] H. Triebel, *Theory of function spaces*, Monographs in Mathematics 78, Birkhauser, 1983.

[60] A. Uchiyama, *A constructive proof of the Fefferman-Stein decomposition of $BMO(\mathbb{R}^n)$*, Acta Math. **148** (1982), 215–241.

[61] J.M. Wilson, *Weighted norm inequalities for the continuous square function*, Trans. Amer. Math. Soc. **314** (1989), 661–692.

[62] P. Wojtaszczyk, *A mathematical introduction to wavelets*, London Mathematical Society Student Texts 37, Cambridge University Press, 1997.

[63] M. Yamazaki, *Navier-Stokes equations in various function spaces* (Japanese), Sugaku **51** (1999), 291–308.

Department of Mathematics, Faculty of Science, Hokkaido University, Sapporo 060-0810, Japan

E-mail address: tachizaw@math.sci.hokudai.ac.jp

Translated by KAZUYA TACHIZAWA

Amer. Math. Soc. Transl.
(2) Vol. **211**, 2003

Recent Topics on Nonlinear Partial Differential Equations: Structure of Radial Solutions for Semilinear Elliptic Equations

Eiji Yanagida and Shoji Yotsutani

For the past two decades, there has been a considerable development in the theory of nonlinear partial differential equations, and many interesting facts have been discovered for various equations. In this article, we will focus on the structure of radial solutions for semilinear elliptic equations with power nonlinearity, and related topics. Our objective in this article is to survey the recent progress as clearly as possible.

Radial solutions play a fundamental role in the investigation of the structure of solutions in semilinear elliptic equations. In radially symmetric solutions, the problems are reduced to the analysis of ordinary differential equations that the radial solutions must satisfy. However, even in this simpler situation, it is not easy to show the existence, non-existence, and uniqueness of solutions through the standard techniques of ordinary differential equations and functional analysis.

Traditionally, these problems have been studied separately, depending on the slight difference of nonlinear terms or the boundary conditions. However, it is gradually becoming clear that this class of equations, which appears to be different, has similar structures for its solutions. Moreover, recent studies have shown that boundary value problems, which radial solutions should satisfy, can be reduced to a certain canonical form by changing variables. By virtue of this, we can understand known results systematically, clarify the unknown structure of various equations of this class, and investigate the structure of radial solutions for semilinear elliptic equations with power nonlinearity more precisely.

In §1 of this article, we briefly summarize results on the radial symmetry of solutions. In §§2–5, we describe the structure of radial solutions to some typical equations. In Section 6, we will survey the methods for the proofs. In §7, we will introduce transformations to reduce the various equations to a canonical form, and clarify the implication of the Kelvin transformation and the Rellich-Pohozaev identity, which play fundamental roles in the analysis of elliptic equations. Concluding remarks are in §8.

This article originally appeared in Japanese in Sūgaku **51** (3) (1999), 276–290.

2000 *Mathematics Subject Classification.* Primary 35J65; Secondary 34A12, 34B15, 35B05, 35B35, 35B60.

1. The Properties of Radially Symmetric Solutions

In 1979, under some general conditions on nonlinear terms, Gidas, Ni, and Nirenberg [**GNN1**] obtained very important results concerning the relation between the symmetry of positive solutions for Dirichlet problems and the symmetry of equations and domains.

For instance, let us consider the Dirichlet problem

$$\Delta u + f(u, |x|) = 0 \text{ in } B, \quad u > 0 \text{ in } B, \quad u = 0 \text{ on } \partial B, \tag{1.1}$$

where $B := \{x \in \mathbf{R}^n; |x| < 1\}$ is a unit ball, and $f(u,r)$ is a smooth function of u and $r = |x|$. It was shown in [**GNN1**] that if $f_r = \partial f/\partial r \leq 0$, then any positive solution of (1.1) must be radially symmetric. In this case the positivity of solutions is essential. Indeed, sign-changing solutions are not necessarily radially symmetric. Also, an example can be shown in which $f_r > 0$ for some r with a positive non-radial solution. Their proof was based on the so-called "moving plane method", first introduced by Alexandroff [**A**], then utilized by Serrin in his paper [**S**].

Gidas, Ni, and Nirenberg [**GNN2**] extended the above result to the case where the domain is $\mathbf{R}^n$, and showed that any positive solution decaying sufficiently fast at ∞ must be radially symmetric. Since then, many results in [**GNN1**] and [**GNN2**] have been improved and proofs simplified. For instance, the fast decay condition on solutions was almost removed by Li and Ni [**LN3**] and Li [**L2**]. Moreover, various refinements and extensions of the moving plane method as well as their applications to a priori estimates are being done extensively (see, e.g., Berestycki, Caffarelli, and Nirenberg [**BCN**], Chen and Lin [**CL2**], and Naito and Suzuki [**NyS**]). Thus, in many cases, the investigation of the structure of positive solutions on a ball or an entire space can be reduced to the structure of positive radial solutions.

We should note that there is a Dirichlet problem on the annulus $A = \{x : a < |x| < b\}$ $(0 < a < b)$ where it has positive solutions, but not necessarily radially symmetric. In fact, we can construct both radially symmetric solutions and non-radially-symmetric solutions in such a case ([**BN**]). This is because we can use the variation method to construct a non-symmetric solution with lower energy than a radially symmetric solution. It is also understandable that a non-symmetric solution can be treated as a special case of a symmetric solution.

It is not always true that the shape of positive solutions of a Neumann problem for semilinear elliptic equations is radially symmetric, even if the domain is the unit ball, and their maxium value may be located at the boundary of the domain. However, there is a strong relationship between the shape of the region and that of solutions on any region, not necessarily the unit ball. For example, Ni and Takagi have shown that solutions with the minimum energy of a scalar field equation will attain the maximum value near a point on the boundary of a given region where the mean curvature becomes maxium ([**N6**]). In this case, it is essential to understand that this solution can be approximated to a radially symmetric solution aound the point where the solution attains the maxium value.

Even for a problem that does not guarantee the symmetry of solutions it is of first importance to study the structure of the sphere-symmetric solutions. Meanwhile, for example, it is important to investigate radially symmetric solutions of equations modeled from astrophysics, since the nature of the equations of astrophysics demands the radially symmetric property.

A radial solution $u = u(|x|)$ of (1.1) satisfies the ordinary differential equation

$$u_{rr} + \frac{n-1}{r} u_r + f(u, r) = 0.$$

Thus the problem is much simplified, and in some cases, much information can be derived by using an elementary method usually found in ODE. However, even if we combine known techniques from the ODE approach, such as shooting methods, and the functional-analytic approach, such as variational methods, it is still not easy to investigate the existence, non-existence, and uniqueness of solutions.

In this article, we shall give a brief survey about the structure of radial solutions of equations with simplest nonlinearity, such as u^p. Even for equations with such simple nonlinearity, the structure of solutions can be very complicated, and thus its analysis is not an easy task. Conversely, under suitable conditions, we can observe the very beautiful structure of solutions, though the implication of the conditions and the mechanism involved are not well understood.

2. The Lane-Emden Equation

First, let us consider the following Dirichlet problem on a bounded domain $\Omega \subset \mathbf{R}^n$:

$$\Delta u + u^p = 0 \text{ in } \Omega, \quad u > 0 \text{ in } \Omega, \quad u = 0 \text{ on } \partial\Omega, \tag{2.1}$$

where $p > 1$ is a parameter. This equation, the Lane-Emden equation (or the Emden-Fowler equation), is a model equation in astrophysics that describes the density distribution of stars. From an application point of view, radial solutions in the case where Ω is a ball are of particular interest [**C**].

It is a little complicated to determine if there is a positive solution for (2.1). If $n = 1$ and 2 with any $p > 1$, or $n \geq 3$ with $1 < p < (n+2)/(n-2)$, there is at least one positive solution on any bounded domain. On the other hand, it has been proved that if $p \geq (n+2)/(n-2)$ with Ω being a domain of star-like shape, there is no positive solution ([**P1**]). Furthermore, it is still a conjecture that if Ω is convex, there is at most one positive solution provided that the equation has solutions.

If $p = 1$, then (2.1) is linear. It is only when there is a bounded solution for this equation that the Laplacian on Ω has the first eigenvalue 1. When $0 < p < 1$, it is possible to show that there is a unique solution on an arbitrary region using the super-subsolution method.

In the following equation, let Ω be a unit ball with its center at the origin. Gidas, Ni, and Nirenberg found that any positive solution must be radially symmetric. Thus, we may investigate the auxiliary initial value problem

$$u_{rr} + \frac{n-1}{r} u_r + u_+^p = 0, \quad u(0) = \alpha > 0, \tag{2.2}$$

where $u_+ = \max\{u, 0\}$. Here, we intentionally removed the initial condition $u_r(0) = 0$, which is usually assumed, since it is a superfluous condition. Indeed, the initial value problem (2.2) has a unique solution, and the condition $u_r(0) = 0$ is automatically satisfied (see, e.g., [**N3**] or [**NY2**]). We denote the solution by $u = u(r; \alpha)$. (In what follows, we use similar notation for solutions of other equations.) We obtain a solution of (2.1) if we find $\alpha^* > 0$ such that $u(1; \alpha^*) = 0$.

For the Lane-Emden equation, it follows from the homogeneity of the nonlinear term that solutions have scaling invariance:

$$u(r;\alpha) \equiv \alpha u(\alpha^{(p-1)/2} r; 1). \tag{2.3}$$

In fact, the following result is known ([**GNN1**]).

THEOREM 2.1. *Suppose that $n \geq 3$. Then the structure of radial solutions of the Lane-Emden equation is as follows:*

(a) *If $1 < p < (n+2)/(n-2)$, then there exists $\alpha^* > 0$ such that*
$u(r;\alpha) > 0$ *on* $[0,1]$ *for* $\alpha \in (0, \alpha^*)$,
$u(r;\alpha) > 0$ *on* $[0,1)$ *and* $u(1;\alpha) = 0$ *for* $\alpha = \alpha^*$,
$u(1;\alpha)$ *has a zero on the interval* $(0,1)$ *for* $\alpha \in (\alpha^*, \infty)$.

(b) *If $p \geq (n+2)/(n-2)$, then $u(r;\alpha) > 0$ on $[0,1]$ for any $\alpha > 0$.*

We note that for $n = 1, 2$, the structure of solutions is as in (a) of the above theorem for any $p > 1$.

It is known that there is a unique radially symmetric positive solution for any $p > 1$ and $n \geq 1$ for the Lane-Emden equation on $A := \{x : a < |x| < b\}(0 < a < b)$ ([**N2**], [**NN**]). However, it is nowhere near so easy to prove the uniqueness of such a solution, because the case of this domain does not hold the scaling invariance equivalent to (2.3).

3. The Matukuma Equation and Conformal Scalar Curvature Equations

We consider the following elliptic equations:

$$\Delta u + K(x) = 0 \text{ in } \mathbf{R}^n. \tag{3.1}$$

Such equations appear in various fields of science. For instance, when $K(x) \equiv 1$, it is the Lane-Emden equation. When $K(x) = (1+|x|^2)^{-1}$, it is called Matukuma's equation. Around 1930, T. Matukuma, an astrophysicist, proposed this model equation to describe the dynamics of a globular cluster of stars, where u represents the gravitational potential. Also, the equation with $K(x) = |x|^{\sigma-2}(1+|x|^2)^{-\sigma/2}$ appears in astrophysics, and is called the Batt-Faltenbacher-Holst equation ([**BFH**]). When $\sigma = 2$, the Batt-Faltenbacher-Holst equation coincides with Matukuma's equation.

If $n \geq 3$ with $p = (n+2)/(n-2)$ for (3.1), it becomes an example of the conformal scalar curvature equation. This is closely related to Yamabe's problem, and studying the existence and non-existence of a positive solution for (3.1) over $\mathbf{R}^n$ is equivalent to studying how to embed conformal Riemann metric and the standard metric into $\mathbf{R}^n$ given that the function $K(x)$ should have a scalar curvature.

A systematic study of the structure of solutions for the equation (3.1) originated from work of Ni [**N1**] in 1982, and many studies have been done since that time. The structure of solutions is slightly different depending on whether or not the decay rate of $K(x)$ at infinity is faster than $|x|^{-2}$. First, the existence and non-existence were established when the decay rate is faster than $|x|^{-2}$ ([**N1**], [**K**], [**N3**], [**KSY**]). Later, Ding and Ni [**DN1**] found a striking phenomenon in the conformal scalar curvature equation: the structure of solutions drastically changes under a slight perturbation of $K(r)$ from $K(r) \equiv$ *positive const*. For Matukuma's equation, the decay order of $K(x)$ is equal to $|x|^{-2}$, so we cannot directly apply the method of analysis to the case where $K(x)$ decays faster than $|x|^{-2}$.

Assuming $K = K(|x|)$, let us consider positive radial solutions. Similarly to (2.2), we investigate the initial value problem

$$u_{rr} + \frac{n-1}{r}u_r + K(r)u_+^p = 0, \quad u(0) = \alpha > 0. \tag{3.2}$$

If $p > 1$ and $K(r) \le 0$, then the solution $u = u(r;\alpha)$ is increasing in α for each fixed $r > 0$. (A complete classification of the structure of all positive solutions of the partial differential equation (3.1) is given in [**CN**].) The case where $K(r) \ge 0$ and $0 < p < 1$ is similar. In these two cases, solutions of (3.2) are in layers parameterized by α, so that the structure is simple. In what follows, we restrict our argument to the case where $K(r) \ge 0$ and $p > 1$.

If $u(r) > 0$ on $[0,\infty)$, we can show that $u(r)$ decays to 0 like a fundamental solution of Δ (we call such a solution a rapid-decay solution), or more slowly than the fundamental solution (we call such a solution a slow-decay solution [**NY2**]). Therefore, if $n \ge 3$, the behavior of solutions $u(r;\alpha)$ of (3.2) as $r \to \infty$ is classified into one of the following three types:

a crossing solution : $u(r)$ has a zero in $(0,\infty)$.

a rapid-decay solution : $u(r) > 0$ on $(0,\infty)$ and $r^{n-2}u$ converges to a finite positive constant as $r \to \infty$.

a slow-decay solution : $u(r) > 0$ on $(0,\infty)$ and $r^{n-2}u(r)$ diverges to ∞ as $r \to \infty$.

When we fix the exponent p and vary $\alpha > 0$ as a parameter, the simplest structure of solutions we first observe are the following three cases, in which the behavior of the solutions is independent of α:

Type C : $u(r;\alpha)$ is a crossing solution for any $\alpha > 0$.

Type R : $u(r;\alpha)$ is a rapid-decay solution for any $\alpha > 0$.

Type S : $u(r;\alpha)$ is a slow-decay solution for any $\alpha > 0$.

If the dimension is $n = 1$ or 2, it is easy to see that any solution of (3.2) must be a crossing solution, which implies that the structure is type C. For $n \ge 3$, the structure depends on $K(|x|)$ and p.

We consider the conformal scalar curvature equation

$$u_{rr} + \frac{n-1}{r}u_r + K(r)u_+^{(n+2)/(n-2)} = 0, \quad u(0) = \alpha > 0. \tag{3.3}$$

When $K(r) \equiv K$ (a positive constant), all solutions are explicitly expressed as

$$u(r;\alpha) = \alpha\left\{1 + \frac{K\alpha^{4/(n-2)}}{n(n-2)}r^2\right\}^{-(n-2)/2},$$

which implies that the solution $u(r;\alpha)$ is a rapid-decay solution for any $\alpha > 0$, so that the structure is of type R. It was shown by Ding and Ni [**DN1**] that the structure changes drastically when monotone perturbations are added to the positive constant $K(r)$. Later, Kusano and Naito [**KN**] extended the result to general p. Furthermore, Yanagida and Yotsutani [**YY1**], [**YY4**] obtained a result including the case of non-monotone perturbations. These results are summarized in the following theorem.

THEOREM 3.1. *Suppose that $K(r) > 0$ is non-constant. Then the structure of solutions of the conformal scalar curvature equation (3.3) is as follows:*

(a) *If $K(r)$ is non-decreasing, the structure is of Type C.*

(b) *If $K(r)$ is non-increasing, the structure is of Type S.*

(c) *If there exists $R \in (0, \infty)$ such that $K(r)$ is non-decreasing on $(0, R)$, non-increasing on (R, ∞), and $\lim_{r\to 0} K(r) = \lim_{r\to\infty} K(r) > 0$, then the structure is of Type M: there exists an $\alpha^* > 0$ such that*
$u(r;\alpha)$ is a crossing solution for $\alpha \in (\alpha^*, \infty)$,
$u(r;\alpha)$ is a rapid-decay solution for $\alpha = \alpha^*$,
$u(r;\alpha)$ is a slow-decay solution for $\alpha \in (0, \alpha^*)$.

This theorem implies that the structure of radial solutions of the conformal scalar curvature is not structurally stable under perturbation from $K(r) \equiv$ positive constant. We can understand its mechanism more clearly by considering the structure in (p, α)-space. If $K(r) \equiv$ positive constant, the type of solutions of (2.3) depends on p only, and is independent of the initial data by virtue of the scaling invariance. In fact, if $1 < p < (n+2)/(n-2)$, all solutions are crossing solutions; whereas if $p > (n+2)/(n-2)$, all solutions are slow-decay solutions. Namely, in (p, α)-space, we see that the set of slow-decay solutions and the set of crossing solutions are separated by the set of rapid-decay solutions, which is a line $p = (n+2)/(n-2)$.

When we add a perturbation to $K(r) \equiv$ positive constant, the structure changes continuously in (p, α)-space, and the set of rapid-decay solutions changes its shape from the line $p = (n+2)/(n-2)$ to a nearby smooth curve. However, if we examine the structural change of solutions only on the line $p = (n+2)/(n-2)$, then the structure appears to change abruptly. Morishita, Yanagida, and Yotsutani [**MYY1**] studied how the set of rapid-decay solutions changes in terms of the perturbation of $K(r)$.

If $K(r)$ does not satisfy the assumptions of Theorem 3.1, the structure of solutions of the conformal scalar curvature equation can be very complicated (see, e.g., [**YY1**], [**BE**], [**ST**]). For instance, it has been shown that there may exist infinitely many rapid-decay solutions (see, e.g., Chen and Lin [**CL3**].

Next, we describe the structure of radial solutions of Matukuma's equation. It follows from the form of $K(r)$ that solutions do not have scaling invariance. Hence the analysis of the structure is not easy. Around 1930, Matukuma conjectured about the structure of positive radial solutions based on his numerical computation. In 1986, Ni and Yotsutani [**NY1**], [**NY2**] started a mathematical investigation of the structure of radial solutions. They proved that Matukuma's conjecture is not true, and developed a new conjecture:

if $1 < p < (n+2)/(n-2)$, then the structure is of Type M,
if $p \geq (n+2)/(n-2)$, then the structure is of Type S.

Later, the radial symmetry of positive solutions, the existence of rapid-decay solutions, and the precise asymptotic behavior of the slow-decay solutions were proved in a series of papers by Li and Ni ([**LN1**], [**LN2**], [**LN3**], [**LN4**]) and Li ([**L1**], [**L2**]). Finally, the above conjecture about the structure of positive radial solutions was confirmed by the proof of uniqueness due to Yanagida [**Y1**].

Concerning the radial symmetry, Li [**L2**] proved that for $p < (n+2)/(n-2)$, any positive solution must be radially symmeric.

These results extend [**GNN2**] to a more general class of equations including (3.1). However, it is not known at present if positive solutions must be radially symmetric for $p > (n+2)/(n-2)$.

We now introduce the structure theorem for the more general $K(r)$ that was motivated by the study of the structure of solutions for the conformal scalar curvature equation and Matukuma's equation. We define functions $G(r)$ and $H(r)$ by

$$G(r) := \frac{1}{p+1} r^n K(r) - \frac{n-2}{2} \int_0^r s^{n-1} K(s) ds,$$
$$H(r) := \frac{1}{p+1} r^{2-(n-2)p} K(r) - \frac{n-2}{2} \int_r^\infty s^{1-(n-2)p} K(s) ds,$$

and we let r_G be the first point at which $G(r_G) < 0$, and r_H the first point from $r = \infty$ at which $H(r_H) < 0$. The functions $G(r)$ and $H(r)$ will appear in the following Rellich-Pohozaev identity:

$$\frac{d}{dr} P(r; u) \equiv G_r(r) u_+^{p+1} = r^{9n-2)(p+1)} H_r(r) u_+^{p+1} \tag{3.4}$$

$$= \frac{1}{p+1} r^{n-1} \left\{ r K_r(r) - \frac{(n-2)p - (n+2)}{2} K(r) \right\} u_+^{p+1}, \tag{3.5}$$

where

$$P(r; u) := \frac{1}{2} r^{n-1} u_r \{ r u_r + (n-2) u \} + \frac{1}{p+1} r^n K(r) u^{p+1}.$$

If $G(r)$ cannot be defined at $r = 0$ due to the singularity of $K(r)$, there is no solution for (3.2). Moreover, if $H(r)$ cannot be defined due to the strong growth rate of $K(r)$ when $r \to \infty$, the structure is of Type C for (3.2). These are understandable from the classical results of Atkinson ([**At**]). If $G(r) \equiv 0$, then $K(r)$ is a constant multiple of $r^{((n-2)p-(n+2))/2}$, and the structure of solutions is of type R.

The following result is based on the work of Yanagida and Yotsutani [**YY1**].

THEOREM 3.2. *Let $n \geq 3$. If $G(r) \not\equiv 0$, then the structure of solutions of (3.2) is as follows:*

(a) *If $r_H = 0$, then the structure is of Type S.*
(b) *If $r_G = \infty$, then the structure is of Type C.*
(c) *If $0 < r_H \leq r_G < \infty$, then the structure is of Type M.*

The above theorem fails if $r_G < r_H$. We can construct a counterexample under this condition. If we impose the condition

$$rK_r(r)/K(r) \text{ is non-increasing on } (0, \infty), \tag{3.6}$$

then the structures of solutions can be analyzed for any p. The following is due to Kawano, Yanagida, and Yotsutani [**KYY1**].

THEOREM 3.3. *Let $n \geq 3$. Suppose that (3.6) holds, and*

$$\ell := \lim_{r \to \infty} rKr(r)/K(r) < \sigma := \lim_{r \to 0} rK_r(r)/K(r).$$

Then the structure of solutions of (3.2) is as follows:

(a) *If $p \leq (n+2+2\ell)/(n-2)$, then the structure is of Type C.*
(b) *If $p \geq (n+2+2\sigma)/(n-2)$, then the structure is of Type S.*
(c) *If $(n+2+2\ell)/(n-2) < p < (n+2+2\sigma)/(n-2)$, then the structure is of Type M. Moreover, the unique initial value $\alpha = \alpha^*(p)$ for which $u(r;\alpha)$ becomes a rapid-decay solution is a C^1-function of $p \in ((n+2+2\ell)/(n-2), (n+2+2\sigma)/(n-2))$.*

It is clear that both Matukuma's equation and the Batt-Faltenbacher-Holst equation satisfy the conditions $r_H \leq r_G$ and (3.6). Hence the structures of solutions for both types of equations can be determined. If the condition (3.6) is not satisfied, it is difficult to analyze the structures of such solutions. This is, for example, because there is no guarantee of the uniqueness of rapid-decay solutions.

Replacing u_+^p with $|u|^{p-1}u$, consider the following equation:

$$u_r r + (n-1)r^{-1}u_r + K(r)|u|^{p-1}u = 0, \quad r > 0, \quad u(0) = \alpha > 0. \tag{3.7}$$

Under the condition (3.6), the structure of all solutions of (3.7) can be determined. The following result is due to Yanagida [**Y3**].

THEOREM 3.4. *Under the assumptions of Theorem 3.3, the structure of solutions of (3.7) is as follows:*

(a) *If* $p \leq (n+2+2\ell)/(n-2)$, *then* $u(r;\alpha)$ *has infinitely many zeros for any* $\alpha > 0$.

(b) *If* $p \geq (n+2+2\sigma)/(n-2)$, *then* $u(r;\alpha)$ *is a slow-decay solution with no zeros for any* $\alpha > 0$.

(c) *If* $(n+2+2\ell)/(n-2) < p < (n+2+2\sigma)/(n-2)$, *then there exists a unique sequence* $0 = \alpha_0 < \alpha_1 < \alpha_2 < \cdots \to \infty$ *such that*

$u(r;\alpha)$ *is a slow-decay solution with exactly* i *zeros for* $\alpha \in (\alpha_i, \alpha_{i+1})$,
$u(r;\alpha)$ *is a rapid-decay solution with exactly* i *zeros for* $\alpha = \alpha_{i+1}$.

4. The Brezis-Nirenberg Equation

Let λ be a positive parameter, and consider the following Dirichlet problem:

$$\Delta u + \lambda u + u_+^{(n+2)/(n-2)} = 0 \text{ in } B, \quad u > 0 \text{ in } B, \quad u = 0 \text{ on } \partial B, \tag{4.1}$$

where $n \geq 3$ and B is a unit ball in $\mathbf{R}^n$. In the case where $\lambda \leq 0$, it can be shown that there is no solution for (4.1), using the Rellich-Pohozaev identity.

In 1983, Brezis and Nirenberg [**BN**] obtained the existence theorem using a new variational method, combining a peculiar property of the equation and the mountain-pass lemma. In particular, for $n = 3$, they clarified the range of λ for the non-existence of solutions by using a variant of the Rellich-Pohozaev identity. The uniqueness of solutions in this case is again a difficult problem, but several years later, Zhang [**Z**] and Kwong and Li [**KL**] proved it.

We can summerize these results in the following theorems. By [**GNN1**], we introduce the following initial value problem:

$$u_{rr} + \frac{n-1}{r}u_r + \lambda u + u_+^{(n+2)/(N-2)} = 0, \quad u(0) = \alpha > 0. \tag{4.2}$$

Let $\lambda_1 > 0$ be the first eigenvalue for $-\Delta$ under Dirichlet boundary conditions on the unit ball. In particular, if $n = 3$, $\lambda_1 = \pi^2$.

When $n \geq 4$, the structure of solutions is rather simple, as the following result shows.

THEOREM 4.1. *Let* $n \geq 4$. *Then the structure of solutions of (4.2) is as follows:*

(a) *If* $0 < \lambda < \lambda_1$, *there exists* $\alpha^* = \alpha^*(\lambda) > 0$ *such that*

$u(r;\alpha) > 0$ *on* $[0,1]$ *for* $\alpha \in (0,\alpha^*)$;
$u(r;\alpha) > 0$ *on* $[0,1)$ *and* $u(1;\alpha) = 0$ *for* $\alpha = \alpha^*$;
$u(r;\alpha)$ *has a zero in* $(0,1)$ *for* $\alpha \in (\alpha^*,\infty)$.

(b) *If* $\lambda \geq \lambda_1$, $u(r;\alpha)$ *has a zero in* $(0,1)$ *for any* $\alpha > 0$.

When $n = 3$, the structure is a little more complicated, and it is called a critical dimension.

THEOREM 4.2. *Let* $n = 3$. *Then the structure of solutions of (4.2) is as follows.*

(a) *If* $0 < \lambda \leq \pi^2/4$, *then* $u(r;\alpha) > 0$ *on* $[0,1]$ *for any* $\alpha > 0$.
(b) *If* $\pi^2/4 < \lambda < \pi^2$, *then there exists* $\alpha^* = \alpha^*(\lambda) > 0$ *such that*
$u(r;\alpha) > 0$ *on* $[0,1]$ *for* $\alpha \in (0,\alpha^*)$;
$u(r;\alpha^*) > 0$ *on* $[0,1)$ *and* $u(1;\alpha^*) = 0$ *for* $\alpha = \alpha^*$;
$u(r;\alpha)$ *has a zero in* $(0,1)$ *for* $\alpha \in (\alpha^*,\infty)$.
(c) *If* $\lambda \geq \pi^2$, *then* $u(r;\alpha)$ *has a zero in* $(0,1)$ *for any* $\alpha > 0$.

Thus, the problem (4.1) has a unique solution if and only if $0 < \lambda < \lambda_1$ for $n \geq 4$ and $\pi^2/4 < \lambda < \pi^2$ for $n = 3$.

Comparing equation (4.2) to (2.2) and (3.2), we observe that the above theorems bear some similarities to Theorems 2.1 and 3.3 with λ corresponding to p.

5. The Scalar Field Equation

The equation

$$\Delta u - u + u^p = 0 \text{ in } \mathbf{R}^n \tag{5.1}$$

is called a scalar field equation; it appears in the study of non-linear Schrödinger equations. We consider the structure of radial solutions of the scalar field equation (5.1). Due to the term $-u$, any solution converging to 0 as $r \to \infty$ must decay exponentially. Hence, the sufficient condition in [**GNN2**] for radial symmetry is satisfied, so that any positive solution converging to 0 must be radially symmetric. The existence and non-existence of positive decaying solutions of (5.1) was shown by Nehari [**Ne**] in 1963. Although the equation looks quite simple, the uniqueness of solutions is a difficult problem. In 1989, Kwong finally obtained a complete proof of the uniqueness, based on the results of Coffman [**Co**] and McLeod and Serrin [**MS**].

It is known that the structure of solutions of the initial value problem

$$u_{rr} + \frac{n-1}{r}u_r - u + u_+^p = 0, \quad u(0) = \alpha > 0, \tag{5.2}$$

is as follows.

THEOREM 5.1. *Let* $n \geq 3$. *Then the structure of solutions of (5.2) is as follows.*

(a) *If* $1 < p < (n+2)/(n-2)$, *there exists* $\alpha^* > 0$ *such that*
$u(r;\alpha) \to 1$ *as* $r \to \infty$ *for* $\alpha \in (0,\alpha^*)$;
$u(r;\alpha) > 0$ *on* $(0,\infty)$ *and* $u(r;\alpha) \to 0$ *as* $r \to \infty$ *for* $\alpha = \alpha^*$;
$u(r;\alpha)$ *has a zero in* $(0,\infty)$ *for* $\alpha \in (\alpha^*,\infty)$.
(b) *If* $p \geq (n+2)/(n-2)$, *then* $u(r;\alpha) > 0$ *on* $(0,\infty)$ *and* $u(r;\alpha) \to 1$ *as* $r \to \infty$ *for any* $\alpha > 0$.

Hence, problem (5.1) has a unique positive decaying solution when $1 < p < (n+2)/(n-2)$, while it does not possess such a solution when $p \geq (n+2)/(n-2)$. Moreover, the structures of such a solution are similar to those of Matukuma's equation. We note that if $n = 1, 2$, then we are always under (a) for any $p > 1$. We don't know if there are results similar to Theorems 3.1–3.4 for the equations

$$u_{rr} + \frac{n-1}{r}u_r - u + K(r)u^p = 0,$$

$$u_{rr} + \frac{n-1}{r}u_r - u + |u|^{p-1}u = 0.$$

6. Existence and Uniqueness of Positive Radial Solutions

As we briefly explained in the previous sections, the existence and uniqueness of (positive) radial solutions of partial differential equations with given boundary conditions or asymptotic behavior have been studied extensively by many researchers since 1980 (see, e.g., [**NPS1**], [**NPS2**], [**N4**], [**N5**], [**LNPS**], [**Su**]).

There are two methods used to establish the existence of solutions. First is the variational method, by which we can find radial solutions of partial differential equations. Since solutions are not necessarily stable in the case of $p > 1$, we need to combine it with the mountain-pass lemma, or constrained minimizing problem (see, e.g., [**N1**], [**GNW**], [**Lo**], [**BN**], [**DN2**], [**R**], [**LnLs**], [**ST**], [**E**]). However, it is hard to show the existence of slow-decay solutions using the variational method, since it is difficult to formulate a suitable function space. Second is a shooting method, from the origin or given boundaries. When the domain is not bounded, we need some special devices, such as topological methods ([**CC**]) or the openness of the set of slow-decay solutions, which is derived by using the Rellich-Pohozaev identity ([**KYY1**], [**YY1**], [**YY4**], [**YY5**]), in order to deal with the asymptotic behavior of solutions at infinity.

Uniqueness of solutions is often an extremely difficult problem. Generally speaking, there are two approaches to the proof. The first is to investigate the properties of $\partial u(r;\alpha)/\partial\alpha$, which is the derivative of the solution $u(r;\alpha)$ with respect to the initial value α. This method has its origin in Kolodner's work [**Ko**]. We study the properties of $\partial u(r;\alpha)/\partial\alpha$ by comparing it with other functions whose properties are easier to derive. To achieve this, we need to find special identities depending on the equation (see, e.g., [**N2**], [**NN**], [**MS**], [**Kw**], [**CL1**], [**Y2**], [**Y3**], [**Y4**], [**Y6**]). The second is to compare a solution satisfying the boundary conditions with other solutions directly through the Rellich-Pohozaev type identities. This method was first used by Yanagida [**Y1**] for the study of Matukuma's equation. Later, this method was developed in [**Y2**], [**KL**], [**YY1**], [**KYY1**], [**ET**], [**KbT**] and other papers.

As we mentioned in the previous section, it is also interesting to investigate sign-changing solutions. Many results have been obtained so far (see, e.g., [**JK**], [**YY3**], [**Y5**]), but the analysis has become more technical. In addition, many studies have been done concerning quasi-linear elliptic equations in which the Laplace operator is replaced by the p-Laplace operator or the mean curvature operator (see, e.g., [**NS**], [**KNY**], [**KYY2**], [**KbYY1**], [**HO**]), but we have omitted the details here.

7. Transformation to Canonical Forms

As we described in the previous section, existence, non-existence, and uniqueness have been investigated separately by various methods, according to equations and boundary conditions. However, the equations appearing in sections 2–5 have the common nonlinear term u^p, and similar structures of solutions. This suggests that there might be a common mechanism behind them. In fact, recent studies have found that the boundary value problems (including third boundary conditions) on a finite or infinite interval for the following second order ordinary differential equation with the nonlinear term u^p:

$$u'' + \alpha(x)u' + \beta(x)u + \gamma(x)u^p = 0, \quad -\infty \leq a < x < b \leq +\infty,$$

can be reduced to canonical forms ([**YY2**], [**KbYY2**], [**YY6**]). This means that we can treat various problems in a unified way, and apply results for one equation to other equations through canonical forms. Moreover, the implication of the Kelvin transformation and the Rellich-Pohozaev identity for the original equation can be understood in a more natural manner through a simple canonical form.

Without going into the details, we can show how, using a change of variables, to reduce the equations

- **(i)** Direchlet problem on the unit ball using the Lane-Emden equation,
- **(ii)** Matukuma equation,
- **(iii)** Direchlet problem on the unit ball using the Brezis-Nirenberg equation,
- **(iv)** scalar filed equation,

to a Direchlet problem of the following type:

$$v_{tt} + k(t)v^p = 0, \quad t \in (0,1). \tag{7.1}$$

We note that information regarding the dimension, domain, boundary condition, or $K(r)$ will be included in $k(r)$. By [**YY2**], the given equation is reducible to

$$g(r)^{-1}\{g(r)u_r\}_r + L(r)v^p = 0 \quad \text{in } (a,b), \tag{7.2}$$

where $g(r)$ and $L(r)$ are positive smooth functions over (a,b). For example, $g(r) = r^{n-1}, a = 0, b = \infty$, and $L(r) = K(r)$ for case (ii), and similarly for case (i). In the cases of (iii) and (iv), by using solutions of

$$\phi_{rr} + \frac{n-1}{r}\phi_r \pm \lambda\phi = 0, \quad \phi(0) = 1,$$

we can eliminate the linear term through a transformation $u \to u\phi$ to reduce the equations to (7.2). For instance, for (iii) with $n = 3$ and $0 < \lambda < \pi^2$, we have $\phi(r) = \sin(\mu r)/(\mu r)$, $g(r) = \sin^2(\mu r)/\mu^2$, $a = 0$, $b = 1$, and

$$L(r) = (\sin^2(\mu r)(\cot(\mu r) - \cot(\mu))/(\mu r))^4$$

with $\mu = \lambda^{1/2}$. Further, changing the variables to

$$v(t) := tu(r), \quad t := \left\{1 + \int_b^r \frac{1}{g(s)}\,ds\right\}^{-1},$$

and putting

$$k(t) = t^{-(3+p)}g(r)^2L(r),$$

we can reduce (7.2) to (7.1).

Note that the function $k(t)$ is smooth on $(0,1)$, but can be divergent at each endpoint. So we need to introduce the following definition to classify the behavior of $v(t)$ at $t = 1$, since $v(t)/(1-t)$ is non-decreasing if $v(t) > 0$:

- **(i)** If $\lim_{t\to 1} v(t)/(1-t)$ exists and is positive, then $v(t)$ is called regular at $t = 1$.
- **(ii)** If $\lim_{t\to 1} v(t)/(1-t) = \infty$, then $v(t)$ is called singular at $t = 1$.

Let us examine the relationship between these definitions and those of the original problems. When $t = 1$ corresponds to the boundary of the unit ball, regular solutions in the above definitions correspond to solutions that satisfy the homogeneous (or, zero) Dirichlet boundary condition. Singular solutions are not Lipschitz continuous at $r = 1$, even if they converge to zero as $r \to 1$, or have positive limit as $r \to 1$. When $t = 1$ corresponds to $r = \infty$, a regular solution corresponds to a rapid-decay solution, and a singular solution corresponds to a slow-decay solution.

The Kelvin transformation $u(r) \to s^{n-2}u(s)$ with $s = 1/r$, which is known to be very useful in the analysis of elliptic equations, exchanges the infinity and the origin. In a canonical form, this transformation corresponds to a simple reflection $v(t) \mapsto v(s)$, $s = 1 - t$, with respect to $t = 1/2$.

On the other hand, in the canonical form (7.1), the Rellich-Pohozaev identity (3.4) given in Section 3 can be expressed as

$$\frac{d}{dt}P(t,v) \equiv G_t(v/t)^{p+1} = H_t\{v/(1-t)\}^{p+1}, \tag{7.3}$$

where

$$\begin{aligned} G_t &:= \frac{1}{p+1}\{(t(1-t))^{(p+3)/2}k(t)\}_t\{t/(1-t)\}^{(p+1)/2}, \\ H_t &:= \frac{1}{p+1}\{(t(1-t))^{(p+3)/2}k(t)\}_t\{(1-t)/t\}^{(p+1)/2}. \end{aligned}$$

Observe that these identities are invariant under the Kelvin transformation. It is known that the identity (3.4) provides an efficient method and sharper results for the problems on an infinite interval, but not always on the unit ball, for example. Considering a finite interval or a boundary condition, we see that an appropriate identity is a rewritten form of (7.3) using the inverse transformation that is reducible ot the canonical form (7.1) by a change of variables. For example, the identity of Rellich-Pohozaev type, when considered for a Brezis-Nirenberg equation in [**BN**], is identical to (7.3).

Moreover, in the canonical form (7.1), the functions $G(t)$ and $H(t)$ become

$$\begin{aligned} G &:= \frac{1}{p+1}t^{p+2}(1-t)k(t) - \frac{1}{2}\int_0^t \tau^{p+1}k(\tau)d\tau, \\ H &:= \frac{1}{p+1}(1-t)^{p+2}tk(t) + \frac{1}{2}\int_t^1 (1-\tau)^{p+1}k(\tau)d\tau, \end{aligned}$$

where $rK_r(r)/K(r)$, $(n+2+2l)/(n-2)$, and $(n+2+2\sigma)/(n-2)$ correspond to $t(1-t)\{\log(k(t)t^{p-1}(1-t)^4\}_t$, $5+2l$, and $5+2\sigma$, respectively, in Theorems 3.3 and 3.4.

Incidentally, the identity (7.3) appears to have a strange form. We may wonder what this identity is. If we look at it from a different direction, it is the usual energy equality. Letting $w(s) := (e^s + e^{-s})v(t)$, $t := (1+e^{-2s})^{-1}$, we see that w should satisfy the equation

$$w_{ss} - w + j(s,w) = 0 \quad \text{on } (-\infty, \infty),$$

where $j(s,w) := (t-t^2)^{3/2}k(t,(t-t^2)^{1/2}w)$. Then the usual energy equality in the differential form is given by

$$\frac{1}{2}\frac{d}{ds}\{w_s^2 - w^2 + 2J(s,w)\} = J_s(s,w)$$

with

$$J(s,w) := \int_0^w j(s,\zeta)d\zeta.$$

This corresponds to the Rellich-Pohozaev type identity (7.3) in the canonical form (7.1). Thus, we may say that (7.3) can be regarded as a natural identity from this viewpoint.

Reducing a given equation to canonical form enables us to study the equations in a systematic way. For example, applying Theorems 3.1 and 3.2, we are able to study the problems of scalar field equations ([**YY2**], [**D**]), the equation of type ([**DH**], [**H**])

$$\Delta u + x \cdot \nabla u + \lambda u + u^p = 0,$$

the Neumann problem and the third boundary problems for a Brezis-Nirenberg equation ([**KbYY2**], [**KbYY3**]), and the equations of Matukuma type (including a singularity at the end point) ([**MYY2**]). We also note that the processes for converting a canonical form involve a series of special functions, and so it becomes very hard to manage each process. So one may or may not be able to get meaningful results.

Though systematic applications are now just getting started, it is expected that we will be able to understand the relation among known results in an organic way, which in turn will lead to progress in the study of each equation.

8. Concluding Remarks

In this article, we described mainly the structures of radial solutions for four types of problems (§§2–5). We conclude that they are all reducible to one problem of the canonical form. We often think that non-linear problems are a collection of individually solved problems. However, we can only unveil the invariance of non-linear structures once we have accumulated individual results for various problems. It is extremely important for us to investigate a special case of each non-linear problem.

We regret that we could not introduce some recent interesting topics of various non-linear equations despite a large number of publications in this area. We think that there must be many classes of non-linear differential equations possessing invariant structures out there. We sincerely hope that this area of study will prosper in the near future.

Acknowledgment. We are grateful to Professor Wei-Ming Ni (University of Minnesota) for his kind advice and encouragement to our research, and useful comments on this article.

References

[A] Alexandroff, A. D., Uniqueness theorem for surfaces in the large, IV, Amer. Math. Soc. Transl. (2) 21 (1962), 412-416.

[At] Atkinson, F. V., On second order nonlinear oscillations, Pacific J. Math. 5 (1955), 643-647.

[BFH] Batt, J., Faltenbacher, W., and Horst. E., Stationary spherically symmetric models in stellar dynamics, Arch. Rational Mech. Anal. 93 (1986), 159-183.

[BCN] Berestycki, H., Caffarelli, L. and Nirenberg, L., Monotonicity for elliptic equations in unbounded domains, Comm. Pure Appl. Math. 50 (1997), 1089-1112.

[BE] Bianchi, G. and Egnell, H., An ODE approach to the equation $\Delta u + Ku^{(n+2)/(n-2)} = 0$ in $\mathbf{R}^n$, Math. Z. 210 (1992), 137-166.

[BN] Brezis, H. and Nirenberg, L., Positive solutions of nonlinear elliptic equations involving critical Sobolev exponents, Comm. Pure Appl. Math. 36 (1983), 437-477.

[C] Chandrasekhar, S., An Introduction to the Study of Stellar Structure, Univ. of Chicago Press, 1939.

[CL1] Chen, C.-C. and Lin, C.-S., Uniqueness of the ground state solutions of $\Delta u + f(u) = 0$ in R^n, $n \geq 3$, Comm. Partial Differential Equations 16 (1991), 1549-1572.

[CL2] Chen, C.-C. and Lin, C.-S., Estimates of the conformal scalar curvature equation via the method of moving planes, Comm. Pure Appl. Math. 50 (1997), 971-1017.

[CL3] Chen, C.-C. and Lin, C.-S., Blowing up with infinite energy of conformal metrics on S^n, Comm. Partial Differential Equations **24** (1999), 785-799.

[CC] Cheng, K.-S. and Chern, J.-L., Existence of positive solutions of some semilinear elliptic equations, J. Differential Equations 98 (1992), 169-180.

[CN] Cheng, K.-S. and Ni, W.-M., On the structure of the conformal scalar curvature equation on R^n, Indiana Univ. Math. J. 41 (1992), 261-278.

[Co] Coffman, C. V., Uniqueness of ground state for $\Delta u - u + u^3 = 0$ and a variational characterization of other solutions, Arch. Rational Mech. Anal. 46 (1972), 12-95.

[D] Dancer, N., The effect of domain shapes on the number of positive solutions of certain nonlinear equations, J. Differential Equations 74 (1988), 120-156.

[Do] Dohmen, C., Structure of positive radial solutions to scalar field equations, Nonlinear waves (Proc. Fourth Math. Soc. Japan Internat. Res. Inst., Sapporo, 1995; R. Agemi et al., editors), GAKUTO Internat. Ser. Math. Sci. Appl., vol. 10, Gakkōtosho, Tokyo, 1997, pp. 39–44.

[DN1] Ding, W.-Y. and Ni, W.-M., On the elliptic equation $\Delta u + Ku^{(n+2)/(n-2)} = 0$ and related topics, Duke Math. J. 52 (1985), 485-506.

[DN2] Ding, W.-Y. and Ni, W.-M., On the existence of positive entire solutions of a semilinear elliptic equation, Arch. Rational Mech. Anal. 91 (1986), 283-308.

[DH] Dohmen, C. and Hirose, M., Structure of positive radial solutions to the Haraux-Weissler equation, Nonlinear Anal. 33 (1998), 51-69.

[ET] Erbe, L. and Tang, M., Uniqueness theorems for positive radial solutions of quasilinear elliptic equations in a ball, J. Differential Equations 138 (1997), 351-379.

[E] Evans, L. C., Partial Differential Equations, American Math. Society, 1998.

[GNN1] Gidas, B., Ni, W.-M. and Nirenberg, L., Symmetry and related properties via the maximum principle, Comm. Math. Phys. 68 (1979), 209-243.

[GNN2] Gidas, B., Ni, W.-M. and Nirenberg, L., Symmetry of positive solutions of nonlinear elliptic equations in $\mathbf{R}^n$, Advances in Math., Supplementary Studies 7A (1981), 369-402.

[GNW] Gui, C., Ni, W.-M. and Wang X, On the stability and instability of positive steady states of a semilinear heat equation in $\mathbf{R}^n$, Comm. Pure Appl. Math. 45 (1992), 1153-1181.

[HO] Hashimoto, T. and Otani, M., Nonexistence of weak solutions of nonlinear elliptic equations in exterior domains, Houston J. Math. 23 (1997), 267–290.

[H] Hirose, M., Structure of positive radial solutions to the Haraux-Weissler equation II, Adv. Math. Sci. Appl. 9 (1999), 473–497.

[JK] Jones, C. and Küpper, T., On infinitely many solutions of a semilinear elliptic equation, SIAM J. Math. Analysis 17 (1986), 803-835.

[KbT] Kabeya, Y. and Tanaka, K., Uniqueness of positive radial solutions of semilinear elliptic equations in $\mathbf{R}^n$ and Séré's non-degeneracy condition, Comm. Partial Differential Equations 24 (1999), 563-598.

[KbYY1] Kabeya, Y., Yanagida, E. and Yotsutani, S., Number of zeros of solutions to singular initial value problems, Tohoku J. Math. 50 (1998), 1-22.

[KbYY2] Kabeya, Y., Yanagida, E. and Yotsutani, S., Global structure of solutions for equations of Brezis-Nirenberg type on the unit ball, Proc. Roy. Soc. Edinburgh Sect. A 131 (2001), 647–665.

[KbYY3] Kabeya, Y., Yanagida, E. and Yotsutani, S., Canonical forms and structure theorems for radial solutions to semilinear elliptic problems, Commun. Pure Appl. Anal. 1 (2002), 85–102.

[K] Kawano, N., On bounded entire solutions of semilinear elliptic equations, Hiroshima Math. J. 14 (1984), 125-158.

[KNY] Kawano, N., Ni, W.-M. and Yotsutani, S., A generalized Pohozaev identity and its applications, J. Math. Soc. Japan 42 (1990), 541-564.

[KSY] Kawano, N., Satsuma, J. and Yotsutani, S., Existence of positive entire solutions of an Emden-type elliptic equation, Funkcial. Ekvac. 31 (1988), 121-145.

[KYY1] Kawano, N., Yanagida, E. and Yotsutani, S., Structure theorems for positive radial solutions to $\Delta u + K(|x|)u^p = 0$ in $\mathbf{R}^n$, Funkcial. Ekvac. 36 (1993), 557-579.

[KYY2] Kawano, N., Yanagida, E. and Yotsutani, S., Structure theorems for positive radial solutions to $\mathrm{div}(|Du|^{m-2}Du) + K(|x|)u^q = 0$ in $\mathbf{R}^n$, J. Math. Soc. Japan 45 (1993), 719-742.

[Ko] Kolodner, I. I., Heavy rotating string—a nonlinear eigenvalue problem, Comm. Pure Appl. Math. 8 (1955), 395-408.

[KN] Kusano, T. and Naito, M., Oscillation theory of entire solutions of second order superlinear elliptic equations, Funkcial. Ekvac. 30 (1987), 269-282.

[Kw] Kwong, M. K. , Uniqueness of positive solutions of $\Delta u - u + u^p = 0$ in $\mathbf{R}^n$, Arch. Rational Mech. Anal. 105 (1989), 243-266.

[KL] Kwong, M.-K. and Li, Y., Uniqueness of radial solutions of semilinear elliptic equations, Trans. Amer. Math. 333 (1992), 339-363.

[L1] Li, Y., Asymptotic behavior of positive solutions of equation $\Delta u + K(|x|)u^p = 0$ in $\mathbf{R}^n$, J. Differential Equations 95 (1992), 304-330.

[L2] Li, Y., On the positive solutions of the Matukuma equation, Duke Math. J. 70 (1993), 575-589.

[LN1] Li, Y. and Ni, W.-M., On conformal scalar curvature equation in $\mathbf{R}^n$, Duke Math. J. 57 (1988), 895-924.

[LN2] Li, Y. and Ni, W.-M., On the existence and symmetry properties of finite total mass solutions of the Matukuma equation, the Eddington equation and their generalizations, Arch. Rational Mech. Anal. 108 (1989), 175-194.

[LN3] Li, Y. and Ni, W.-M., On the asymptotic behavior and radial symmetry of positive solutions of semi-linear elliptic equations in $\mathbf{R}^n$, I, Asymptotic behavior, Arch. Rational Mech. Anal. 118 (1992), 195-222.

[LN4] Li, Y. and Ni, W.-M., On the asymptotic behavior and radial symmetry of positive solutions of semi-linear elliptic equations in Rn, II, Radial symmetry, Arch. Rational Mech. Anal. 118 (1992), 223-243.

[LN5] Li, Y. and Ni, W.-M., Radial symmetry of positive solutions of nonlinear elliptic equations in $\mathbf{R}^n$, Comm. Partial Differential Equations 18 (1993), 1043-1054.

[Ln] Lin, C.-S., Uniqueness of least energy solutions to a semilinear elliptic equation in $\mathbf{R}^2$, Manuscripta Math. 84 (1994), 13–19.

[LnLs] Lin, C.-S. and Lin, S.-S., Positive radial solutions for $\Delta u + Ku^{(n+2)/(n-2)} = 0$ in R^n, and related topics, Applicable Anal. 38 (1990), 121-159.

[Lo] Lions, P.-L., On the existence of positive solutions of semilinear elliptic equations, SIAM Review 24 (1982), 441-467.

[LNPS] Lloyd, N. G., Ni, W.-M., Peletier, L. A. and Serrin J. (eds.), Nonlinear Diffusion Equations and Their Equilibrium States 3, Birkhäuser, Boston, 1992.

[MS] McLeod, K. and Serrin, J., Uniqueness of positive radial solutions of $\Delta u + f(u) = 0$ in R^n, Arch. Rational Mech. Anal. 99 (1987), 115-145.

[MYY1] Morishita, H., Yanagida, E. and Yotsutani, S., Structural change of solutions for a scalar curvature equation, Differential and Integral Equations 14 (2001), 273–288.

[MYY2] Morishita, H., Yanagida, E. and Yotsutani, S., Structure of positive radial solutions including all singular solutions, in preparation.

[Nm] Naito, M., A note on bonded positive entire solutions of semilinear elliptic equations, Hiroshima Math. J. 14 (1984), 211-214.

[NyS] Naito, Y. and Suzuki, T., Radial symmetry of positive solutions for semilinear elliptic equations on the unit ball in R^n, Funkcial. Ekvac. 41 (1998), 215-234.

[Ne] Nehari, Z., On a class of nonlinear second order differential equations, Trans. Amer. Math. Soc. 95 (1960), 101-123.

[N1] Ni, W.-M., On the elliptic equation $\Delta u + K(x)u^{(n+2)/(n-2)} = 0$, its generalization, and applications in geometry, Indiana Univ. Math. J. 31 (1982), 493-529.

[N2] Ni, W.-M., Uniqueness of solutions of nonlinear Dirichlet problems, J. Diff. Equations 50 (1983), 289-304.

[N3] Ni, W.-M., Uniqueness, nonuniqueness and related questions of nonlinear elliptic and parabolic equations, Proc. Sympos. Pure Math. 45 (1986), 229-241.

[N4] Ni, W.-M., Some aspects of semilinear elliptic equations on $\mathbb{R}^n$, Nonlinear Diffusion Equations and Their Equilibrium States, II (Berkeley, CA, 1986; W. M. Ni et al., editors), Math. Sci. Res. Inst. Publ., vol. 13, Springer-Verlag, New York, 1988, pp. 171–205.

[N5] Ni, W.-M., Recent progress in semilinear elliptic equations, RIMS Kokyuroku 679 (Kyoto Univ.), (1989), 1-39.

[N6] Ni,W.-M., Diffusion, cross-diffusion, and their spike-layer steady states, Notices Amer. Math. Soc. 45 (1998), 9-18.

[NN] Ni, W.-M. and Nussbaum, R.D., Uniqueness and nonuniqueness for positive radial solutions of $\Delta u + f(u,r) = 0$, Comm. Pure Appl. Math. 38 (1985), 67-108.

[NPS1] Ni, W.-M., Peletier, L. A. and Serrin, J. (eds.), Nonlinear Diffusion Equations and Their Equilibrium States I, Springer-Verlag, New York, 1988.

[NPS2] Ni, W.-M., Peletier, L. A. and Serrin, J. (eds.), Nonlinear Diffusion Equations and Their Equilibrium States II, Springer-Verlag, New York, 1988.

[NS] Ni, W.-M. and Serrin, J., Non-existence theorems for singular solutions of quasi-linear partial differential equations, Comm. Pure Appl. Math. 39 (1986), 379-399.

[NY1] Ni, W.-M. and Yotsutani, S., On Matukuma's equation and related topics, Proc. Japan Acad. Ser. A 62 (1986), 260-263.

[NY2] Ni, W.-M. and Yotsutani, S., Semilinear elliptic equations of Matukuma-type and related topics, Japan J. Appl. Math. 5 (1988), 1-32.

[P1] Pohozaev, S. I., Eigenfunctions of the equation $\Delta u + \lambda f(u) = 0$, Soviet Math. Dokl. 6 (1965), 1408-1411.

[P2] Pohozaev, S. I., On the eigenfunctions of quasilinear elliptic problems. Math. USSR Sb. 11 (1970), 171-188.

[PS] Pucci, P. and Serrin, J., A general variational identity, Indiana Univ. Math. J. 35 (1985), 681-703.

[R] Rabinowitz, P. H., Minimax Method in Critical Point Theory with Applications to Differential Equations, American Mathematical Society, Providence, 1986.

[Re] Rellich, F., Darstellung der Eigenwerte von $\Delta u + \lambda u = 0$ durch ein Randintegral, Math. Z. 46 (1940), 635-636.

[S] Serrin, J., A symmetry problem in potential theory, Arch. Rational Mech. Anal. 43 (1971), 304-318.

[ST] Sasahara, Y. and Tanaka, K., A remark on positive radial solutions of the elliptic equation $\Delta u + K(|x|)u^{(n+2)/(n-2)} = 0$ in $\mathbf{R}^n$, Proc. Amer. Math. Soc. 123 (1995), 527-531.

[Su] Suzuki, T., Semilinear Elliptic Equations, GakkoTosho, Tokyo, 1994.

[Y1] Yanagida, E., Structure of positive radial solutions of Matukuma's equation, Japan J. Indust. Appl. Math. 8 (1991), 165-173.

[Y2] Yanagida, E., Uniqueness of positive radial solutions of $\Delta u + g(r)u + h(r)u^p = 0$ in $\mathbf{R}^n$, Arch. Rational Mech. Anal. 115 (1991), 257-274.

[Y3] Yanagida, E., Uniqueness of positive radial solutions of $\Delta u + f(u,|x|) = 0$, Nonlinear Anal. 19 (1992), 1143-1154.

[Y4] Yanagida, E., Sturmian theory for a class of nonlinear second-order differential equations, J. Math. Anal. Appl. 187 (1994), 650-662.

[Y5] Yanagida, E., Structure of radial solutions to $\Delta u + K(|x|)|u|^{p-1}u = 0$ in $\mathbf{R}^n$, SIAM J. Math. Anal. 27 (1996), 997-1014.

[Y6] Yanagida, E., Uniqueness of rapidly decaying solutions to the Haraux-Weissler equation, J. Differential Equations 127(1996), 561-570.

[YY1] Yanagida, E. and Yotsutani, S., Classifications of the structure of positive radial solutions to $\Delta u + K(|x|)u^p = 0$ in $\mathbf{R}^n$, Arch. Rational Mech. Anal. 124 (1993), 239-259.

[YY2] Yanagida, E. and Yotsutani, S., Pohozaev identity and its applications, RIMS Kokyuroku 834 (Kyoto Univ.) (1993), 80-90.

[YY3] Yanagida, E. and Yotsutani, S., Existence of nodal fast-decay solutions to $\Delta u + K(|x|)|u|^{p-1}u = 0$ in $\mathbf{R}^n$, Nonlinear Anal. 22 (1994), 1005-1015.

[YY4] Yanagida, E. and Yotsutani, S., Existence of positive radial solutions to $\Delta u + K(|x|)u^p = 0$ in $\mathbf{R}^n$, J. Differential Equations 115 (1995), 477-502.

[YY5] Yanagida, E. and Yotsutani, S., Global structure of positive solutions to Matukuma-type equations, Arch. Rational Mech. Anal. 134 (1996), 199-226.

[YY6] Yanagida, E. and Yotsutani, S., A unified approach to the structure of radial solutions for semilinear elliptic problems, Japan J. Indust. Appl. Math. 18 (2001), 503–519.

[Z] Zhang, L., Uniqueness of positive solutions of $\Delta u + u + u^p = 0$ in a ball, Comm. Partial Differential Equations 17 (1992), 1141-1164.

MATHEMATICAL INSTITUTE, TOHOKU UNIVERSITY, SENDAI 980-8578, JAPAN

FACULTY OF SCIENCE AND ENGINEERING, RYUKOKU UNIVERSITY, SETA, OTSU 520-2194, JAPAN

Translated by TATSUHIKO TABARA

Titles in This Series

211 **K. Nomizu, Editor,** Selected papers on analysis and differential equations
210 **S. G. Gindikin, Editor,** Lie groups and symmetric spaces. In memory of F. I. Karpelevich
209 **N. N. Uraltseva, Editor,** Proceedings of the St. Petersburg Mathematical Society, Volume IX
208 **M. V. Karasev, Editor,** Asymptotic methods for wave and quantum problems
207 **Yu. M. Suhov, Editor,** Analytic methods in applied probability
206 **M. A. Shubin and M. S. Agranovich, Editors,** Partial differential equations (Mark Viskik's Seminar)
205 **N. N. Uraltseva, Editor,** Proceedings of the St. Petersburg Mathematical Society, Volume VIII
204 **K. Nomizu, Editor,** Selected papers on classical analysis
203 **N. N. Uraltseva, Editor,** Proceedings of the St. Petersburg Mathematical Society, Volume VII
202 **V. Turaev and A. Vershik, Editors,** Topology, ergodic theory, real algebraic geometry. Rokhlin's memorial
201 **Michael Semenov-Tian-Shansky, Editor,** L. D. Faddeev's seminar on mathematical physics
200 **L. Lerman, G. Polotovskiĭ, and L. Shilnikov, Editors,** Methods of qualitative theory of differential equations and related topics
199 **N. N. Uraltseva, Editor,** Proceedings of the St. Petersburg Mathematical Society, Volume VI
198 **R. A. Minlos, Senya Shlosman, and Yu. M. Suhov, Editors,** On Dobrushin's way. From probability theory to statistical physics
197 **Vladimir Arnold, Maxim Kontsevich, and Anton Zorich, Editors,** Pseudoperiodic topology
196 **Ya. Eliashberg, D. Fuchs, T. Ratiu, and A. Weinstein, Editors,** Northern California symplectic geometry seminar
195 **O. V. Belegradek et al.,** Model Theory and Applications
194 **A. Astashkevich and S. Tabachnikov, Editors,** Differential Topology, Infinite-Dimensional Lie Algebras, and Applications (D. B. Fuchs' 60th Anniversary Collection)
193 **N. N. Uraltseva, Editor,** Proceedings of the St. Petersburg Mathematical Society, Volume V
192 **Lev Beklemishev, Mati Pentus, and Nikolai Vereshchagin,** Provability, Complexity, Grammars
191 **A. Yu. Morozov and M. A. Olshanetsky, Editors,** Moscow Seminar in Mathematical Physics
190 **S. Tabachnikov, Editor,** Differential and Symplectic Topology of Knots and Curves
189 **V. Buslaev, M. Solomyak, and D. Yafaev, Editors,** Differential Operators and Spectral Theory (M. Sh. Birman's 70th anniversary collection)
188 **O. A. Ladyzhenskaya, Editor,** Proceedings of the St. Petersburg Mathematical Society, Volume IV
187 **M. V. Karasev, Editor,** Coherent Transform, Quantization, and Poisson Geometry
186 **A. Khovanskiĭ, A. Varchenko, and V. Vassiliev, Editors,** Geometry of Differential Equations
185 **B. Feigin and V. Vassiliev, Editors,** Topics in Quantum Groups and Finite-Type Invariants (Mathematics at the Independent University of Moscow)
184 **Peter Kuchment and Vladimir Lin, Editors,** Voronezh Winter Mathematical Schools (Dedicated to Selim Krein)

TITLES IN THIS SERIES

183 **K. Nomizu, Editor,** Selected Papers on Harmonic Analysis, Groups, and Invariants

182 **V. E. Zakharov, Editor,** Nonlinear Waves and Weak Turbulence

181 **G. I. Olshanski, Editor,** Kirillov's Seminar on Representation Theory

180 **A. Khovanskiĭ, A. Varchenko, and V. Vassiliev, Editors,** Topics in Singularity Theory

179 **V. M. Buchstaber and S. P. Novikov, Editors,** Solitons, Geometry, and Topology: On the Crossroad

178 **V. Kreinovich and G. Mints, Editors,** Problems of Reducing the Exhaustive Search

177 **R. L. Dobrushin, R. A. Minlos, M. A. Shubin, and A. M. Vershik, Editors,** Topics in Statistical and Theoretical Physics (F. A. Berezin Memorial Volume)

176 **E. V. Shikin, Editor,** Some Questions of Differential Geometry in the Large

175 **R. L. Dobrushin, R. A. Minlos, M. A. Shubin, and A. M. Vershik, Editors,** Contemporary Mathematical Physics (F. A. Berezin Memorial Volume)

174 **A. A. Bolibruch, A. S. Merkur′ev, and N. Yu. Netsvetaev, Editors,** Mathematics in St. Petersburg

173 **V. Kharlamov, A. Korchagin, G. Polotovskiĭ, and O. Viro, Editors,** Topology of Real Algebraic Varieties and Related Topics

172 **K. Nomizu, Editor,** Selected Papers on Number Theory and Algebraic Geometry

171 **L. A. Bunimovich, B. M. Gurevich, and Ya. B. Pesin, Editors,** Sinai's Moscow Seminar on Dynamical Systems

170 **S. P. Novikov, Editor,** Topics in Topology and Mathematical Physics

169 **S. G. Gindikin and E. B. Vinberg, Editors,** Lie Groups and Lie Algebras: E. B. Dynkin's Seminar

168 **V. V. Kozlov, Editor,** Dynamical Systems in Classical Mechanics

167 **V. V. Lychagin, Editor,** The Interplay between Differential Geometry and Differential Equations

166 **O. A. Ladyzhenskaya, Editor,** Proceedings of the St. Petersburg Mathematical Society, Volume III

165 **Yu. Ilyashenko and S. Yakovenko, Editors,** Concerning the Hilbert 16th Problem

164 **N. N. Uraltseva, Editor,** Nonlinear Evolution Equations

163 **L. A. Bokut′, M. Hazewinkel, and Yu. G. Reshetnyak, Editors,** Third Siberian School "Algebra and Analysis"

162 **S. G. Gindikin, Editor,** Applied Problems of Radon Transform

161 **K. Nomizu, Editor,** Selected Papers on Analysis, Probability, and Statistics

160 **K. Nomizu, Editor,** Selected Papers on Number Theory, Algebraic Geometry, and Differential Geometry

159 **O. A. Ladyzhenskaya, Editor,** Proceedings of the St. Petersburg Mathematical Society, Volume II

158 **A. K. Kelmans, Editor,** Selected Topics in Discrete Mathematics: Proceedings of the Moscow Discrete Mathematics Seminar 1972–1990

157 **M. Sh. Birman, Editor,** Wave Propagation. Scattering Theory

156 **V. N. Gerasimov, N. G. Nesterenko, and A. I. Valitskas,** Three Papers on Algebras and Their Representations

155 **O. A. Ladyzhenskaya and A. M. Vershik, Editors,** Proceedings of the St. Petersburg Mathematical Society, Volume I

For a complete list of titles in this series, visit the
AMS Bookstore at **www.ams.org/bookstore/**.